Roger Herger

Perovskite Thin Films

Roger Herger

Perovskite Thin Films

Surface Structure, Interface Structure, and Film Growth

Südwestdeutscher Verlag für Hochschulschriften

Impressum/Imprint (nur für Deutschland/ only for Germany)
Bibliografische Information der Deutschen Nationalbibliothek: Die Deutsche Nationalbibliothek verzeichnet diese Publikation in der Deutschen Nationalbibliografie; detaillierte bibliografische Daten sind im Internet über http://dnb.d-nb.de abrufbar.
Alle in diesem Buch genannten Marken und Produktnamen unterliegen warenzeichen-, marken- oder patentrechtlichem Schutz bzw. sind Warenzeichen oder eingetragene Warenzeichen der jeweiligen Inhaber. Die Wiedergabe von Marken, Produktnamen, Gebrauchsnamen, Handelsnamen, Warenbezeichnungen u.s.w. in diesem Werk berechtigt auch ohne besondere Kennzeichnung nicht zu der Annahme, dass solche Namen im Sinne der Warenzeichen- und Markenschutzgesetzgebung als frei zu betrachten wären und daher von jedermann benutzt werden dürften.

Verlag: Südwestdeutscher Verlag für Hochschulschriften Aktiengesellschaft & Co. KG
Dudweiler Landstr. 99, 66123 Saarbrücken, Deutschland
Telefon +49 681 37 20 271-1, Telefax +49 681 37 20 271-0, Email: info@svh-verlag.de
Zugl.: Zürich, Universität, Dissertation, 2008

Herstellung in Deutschland:
Schaltungsdienst Lange o.H.G., Berlin
Books on Demand GmbH, Norderstedt
Reha GmbH, Saarbrücken
Amazon Distribution GmbH, Leipzig
ISBN: 978-3-8381-0446-1

Imprint (only for USA, GB)
Bibliographic information published by the Deutsche Nationalbibliothek: The Deutsche Nationalbibliothek lists this publication in the Deutsche Nationalbibliografie; detailed bibliographic data are available in the Internet at http://dnb.d-nb.de.
Any brand names and product names mentioned in this book are subject to trademark, brand or patent protection and are trademarks or registered trademarks of their respective holders. The use of brand names, product names, common names, trade names, product descriptions etc. even without a particular marking in this works is in no way to be construed to mean that such names may be regarded as unrestricted in respect of trademark and brand protection legislation and could thus be used by anyone.

Publisher:
Südwestdeutscher Verlag für Hochschulschriften Aktiengesellschaft & Co. KG
Dudweiler Landstr. 99, 66123 Saarbrücken, Germany
Phone +49 681 37 20 271-1, Fax +49 681 37 20 271-0, Email: info@svh-verlag.de

Copyright © 2009 by the author and Südwestdeutscher Verlag für Hochschulschriften Aktiengesellschaft & Co. KG and licensors
All rights reserved. Saarbrücken 2009

Printed in the U.S.A.
Printed in the U.K. by (see last page)
ISBN: 978-3-8381-0446-1

Die vorliegende Arbeit wurde von der Mathematisch-naturwissenschaftlichen Fakultät der Universität Zürich im Frühjahrssemester 2008 als Dissertation unter dem Originaltitel *Surface and Interface Structure and Thin Film Growth of Perovskites* angenommen.

Promotionskomitee
Prof. Dr. Hugo Keller (Vorsitz)
Prof. Dr. Philip R. Willmott (Leitung der Dissertation)
Prof. Dr. Bruce D. Patterson
Dr. Christophe P. Rossel

Meiner Familie

Deus in minimis maximus

Aurelius Augustinus (354 – 430)

Acknowledgments

I would like to thank the many people who have made this work possible and helped create an inspiring, constructive and pleasant working atmosphere.

First, I want to thank Prof. Dr. Philip R. Willmott. It was he who mainly guided me through this thesis. I am indebted to him for his always constructive feedback, his thorough scientific knowledge and his experience to keep the goal of this work in view. Above all, I thank him for his friendship and his good portion of British humor.

My special gratitude belongs to Prof. Dr. Bruce D. Patterson, as a scientist and as a person. His open-minded attitude to always go a step further than originally thought, did impress me. I am modeling myself on his winsome character and his endless motivation.

I would also like to thank Prof. Dr. H. Keller of the Physics Institute of the University of Zürich, who gave me the chance to do my Ph.D. in physics by taking over the official supervision.

I am thankful to Dr. Christophe P. Rossel from IBM Rüschlikon, Zürich, for his guiding advice, especially in the beginning of this work, and for adding an industrial view on this project.

I want to sincerely thank my co-students in the Surface Diffraction group, Christian Schlepütz, Domenico Martoccia, Stephan Pauli, and Matts Björck for all their assistance during the experiments and their helping hands whenever needed, but also for countless scientific discussions that resulted in pushing the Surface Diffraction station forward. Our beamtimes will occupy a special place in my memory.

My special thanks go to Dr. Oliver Bunk for introducing me in the mysteries of FIT and his tremendous efforts to help me solve the surface of strontium titanate.

This work would not have been possible without the development of the Pilatus pixel detectors. Thus, I thank the detector group, Christian Brönnimann, Beat Henrich, Philipp Kraft, Gregor Hülsen, Eric Eikenberry and Petr Salficky for the fruitful collaboration and the outstanding support.

Good physics is always a team-work between experiment and theory. I am hence very grateful to Dr. Bernard Delley of the Condensed Matter Theory Group at the Paul Scherrer Institut for performing density functional theory calculations and his willingness to start up this new collaboration.

I extend my best thanks also to the group of Prof. Dr. Dilano K. Saldin and his co-workers Prof. Dr. Paul F. Lyman and Dr. Valentin L. Shneerson from the Department of Physics and the Laboratory of Surface Studies, University of Wisconsin, Milwaukee, for their help in solving the surface of strontium titanate with PARADIGM.

I am grateful to Prof. Dr. Roy Clarke and Divine Kumah from the Randall Laboratory of Physics, University of Michigan, Ann Arbor, and Prof. Dr. em. Yizhak Yacoby of the Racah Institute of Physics, Hebrew University, Jerusalem, for the fruitful collaboration on the lanthanum strontium manganate work. The extraordinary COBRA phase-retrieval method was invaluable for the structure determination of the thin films.

A fruitful collaboration on lanthanum strontium manganite thin films could also be established with the Surface and Interface Spectroscopy Beamline at the Swiss Light Source. I am thus indebted to Dr. Luc Pathey, Dr. Juraj Krempaský, Dr. Ming Shi, and Dr. Mihaela Falub.

The Rutherford backscattering measurements were carried out by Dr. Max Döbeli from Ion Beam Physics, Paul Scherrer Institut and ETH-Zürich. I thank him for his efforts and the good chats we had on the results.

My special thanks also belong our technicians Dominik Meister and Michael Lange who helped me rapidly solve mechanical and electrical problems as well as the rest of the Materials Science group – Dr. Fabia Gozzo, Dr. Bernd Schmitt, Dr. David Maden and Dr. Marco Stampanoni – for the smooth work at the Beamline.

Further I would like to thank many others from the Paul Scherrer Institut for their help in laboratory-based x-ray diffraction and sample preparation – Prof. Dr. Hans K. Grimmer, PD Dr. Kazimierz Conder, Dr. Ekaterina Pomjakushina and Dr. Denis Cheptiakov.

A substantial part of the characterization work has been carried out by resistivity measurements and atomic force microscopy. I thus thank Stephen Weyeneth, Stela Canulescu, Dr. Slawomir Czekaj and Rolf Schelldorfer for their assistance and interesting discussions.

During my studies at the University of Zürich and at the Paul Scherrer Institut, I perceived the generous support of my parents who never stopped trusting in me. Thank you both!

My very last thank you belongs to my beloved Silvia. I thank her for all her interest in my work, the patience when I had beamtime over the weekend once again, and the freedom to develop myself besides her. Thank you for making myself at home in Liechtenstein!

Abstract

Transition metal oxides with the perovskite structure exhibit fascinating physical properties such as high-temperature superconductivity or colossal magnetoresistance, to name only two of the most prominent examples. Their potential for technological applications cannot be overstated. The flipside of the drive for further miniaturization ("downsizing") is the effect of surface and/or interface modifications of the crystalline structure and consequently on the physical properties of these systems. On the one hand, surface and interface effects can set lower limits to devices that exploit bulk effects, while on the other, new and unexpected phenomena may occur here, that are not evident in bulk materials. The exact knowledge of atomic positions is therefore crucial for the design of nanoscaled devices and may help explaining unexpected physical effects at surfaces and interfaces.

The unique combination of (a) highly brilliant x-rays produced in a modern third generation synchrotron source, (b) the availability of a fast, single-photon counting area pixel detector and (c) a pulsed laser deposition equipment for *in-situ* growth enables one to study both the structure and kinetics of the thin film growth of perovskites.

Surface x-ray diffraction was used as the primary research tool to determine the surface and interface structures strontium titanate ($SrTiO_3$, STO) and thin films of lanthanum strontium manganate ($La_{1-x}Sr_xMnO_3$, LSMO). The structures were refined using conventional minimization of a goodness of fit criterion with the program FIT. Crucial to the structure refinement became the use of the two phase retrieval-methods PARADIGM (in the case of STO) and COBRA (LSMO), as well as extensive density functional theory (DFT) calculations. Further characterization was carried out via laboratory-based x-ray diffraction, atomic force microscopy, x-ray photoelectron spectroscopy, low-energy ion-scattering, reflection high-energy electron diffraction, Rutherford backscattering, and resistivity measurements using the four-point method.

The surface of STO was analyzed for two different environments: One (cold) was at room temperature and in ultra-high vacuum, and the other (hot) at elevated temperatures and in an oxygen background, i.e., under conditions typical for perovskite thin film growth. The cold surface structure comprised of a weighted mixture of a (1×1) relaxation and (2×1) and (2×2) reconstructions, simultaneously present at the surface. The structures were best modeled by a TiO_2-rich surface, where a double TiO_2-layer was present at the surface. The reconstructions were energetically favorable according to DFT. They disappeared within several minutes upon heating to the hot conditions, forming a termination very similar to the cold (1×1), but more puckered and higher in energy. Results using PARADIGM confirmed the TiO_2-rich termination of the hot surface. Surface energy considerations suggested a temperature-induced order-disorder transition, produced by a mixing of the (2×1) and (2×2) reconstructions, to form the pseudo (1×1) structure. Atomic displacements were significant down to three unit cells.

In-situ kinetic studies of the growth of LSMO thin films on STO using pulsed laser deposition (PLD) led to the proposition of a novel energetic smoothing mechanism in the growth of complex metal-oxide thin films. Below 50% monolayer (ML) coverage, prompt insertion of the energetic impinging species into small-diameter islands causes them to break up to form daughter islands. This smoothing mechanism therefore inhibits the formation of large-diameter 2D-islands and the seeding of 3D-growth. Above 50% coverage, islands begin to coalesce and their breakup is thereby suppressed. The energy of the incident flux is instead rechanneled into enhanced surface diffusion.

Structures of six ML-by-ML PLD-grown ultra thin LSMO films with thicknesses between 1 and 9 MLs were determined using the COBRA phase-retrieval method and subsequent structural refinement. Four important results were found. First, the out-of-plane lattice constant is elongated across the substrate-film interface. Second, the transition from substrate to film is not abrupt, but changes gradually over approximately three unit cells. Thirdly, Sr segregates towards the topmost monolayer of the film. We determined a Sr-segregation enthalpy of -15 kJ/mol from the occupation parameters. Finally, the electronic bandwidth W was used to explain the onset of magnetoresistance for films of 9 or more monolayers thickness. Resistivity measurements of the 9 monolayer-thick film confirm magnetoresistance and the presence of a dead layer with mostly insulating properties.

Zusammenfassung

Bei Übergangsmetalloxiden mit Perowskitstruktur treten faszinierende physikalische Effekte auf. Zwei bekannte Beispiele hierzu sind die Hochtemperatursupraleitung und der kolossale Magnetwiderstand. Solche Eigenschaften sind für technologische Anwendungen von enormer Bedeutung. Der Trend zur fortschreitenden Miniaturisierung führt dazu, dass Effekte an Ober- und/oder Grenzflächen eine Veränderung der kristallinen Struktur zur Folge haben, was sich auch auf die physikalischen Effekte dieser Systeme auswirkt. Obwohl der Effekt im Volumenfestkörper zu beobachten ist, können so Oberflächen und Grenzflächen der Miniaturisierung in der technologischen Applikation zum einen untere Grenzen setzen. Zum anderen kann gerade das Vorhandensein einer Grenzfläche dort neue, unbekannte Effekte erzeugen. Die Kenntnis der genauen Atompositionen bildet daher eine wertvolle Grundlage für die Entwicklung von Bausteinen auf der Nanometerskala. Sie kann aber auch bei der Erklärung von solch neuartigen physikalischen Effekten an Oberflächen und Grenzflächen hilfreich sein.

Die einzigartige Kombination von hoch-brillanter Röntgenstrahlung einer modernen Synchrotronquelle dritter Generation, die Verfügbarkeit eines neuartigen und schnellen Flächenzählers mit Einzelphotonendetektion, sowie eine Apparatur zur gepulsten Laser Abscheidung ermöglichen *in-situ* das struktur- und zeitaufgelöste Studium des Wachstums von dünnen Perowskitschichten.

In erster Linie wurde Oberflächenröntgenbeugung zur Strukturbestimmung von Oberflächen und Grenzflächen von Strontiumtitanat ($SrTiO_3$, STO) und dünnen Schichten von Lanthanstrontiummanganat ($La_{1-x}Sr_xMnO_3$, LSMO) verwendet. Die Strukturanpassung mit dem Programm FIT erfolgte durch die Minimierung des Gütefaktors. Von zentraler Bedeutung für die Strukturbestimmung waren die beiden Phasenbestimmungs-Algorithmen PARADIGM (im Falle von STO) und COBRA (bei LSMO), sowie Dichtefunktionalrechnungen. Zur weiteren Charakterisierung wurden durchgeführt: Röntgenbeugung mit einer Laborquelle, Rasterkraftmikroskopie, Photoelektronen-Spektroskopie mit Röntgenstrahlung, Nieder-Energie Ionen-Streuung, Reflexions-Hochenergie-Elektronenbeugung, Rutherford Rückstreuung und Messung des spezifischen Widerstands über die Vierspitzenemethode.

Die Oberfläche von STO wurde unter zwei verschiedenen Umgebungsbedingungen analysiert: Einerseits bei Raumtemperatur und in Ultrahochvakuum (kalt) und andererseits bei hoher Temperatur und in einem Sauerstoffhintergrund (heiss), d.h. bei Bedingungen die typisch sind für das Dünnschicht-Wachstum von Perowskiten. Die Oberfläche des kalten Kristalls zeigte gleichzeitig eine gewichtete Mischung einer (1×1) Relaxation und den beiden Überstrukturen (2×1) und (2×2). Die Strukturen konnten am besten durch eine TiO_2-reiche Oberfläche beschrieben werden, die mit einer Doppellage bestehend aus zwei TiO_2-Schichten terminiert war. Dichtefunktionalrechnung wies die beiden Rekonstruktionen als energetisch günstig aus. Die Überstrukturen verschwanden beim Aufheizen innerhalb von Minuten. Dabei bildete sich eine Oberfläche ähnlich der der kalten (1×1) Struktur, nun jedoch stärker verzerrt und bei höherer Oberflächenenergie. Die TiO_2-reiche Oberfläche der heissen Terminierung wurde durch PARADIGM bestätigt. Betrachtungen der Oberflächenenergie liessen auf einen Übergang von der geordneten zu einer vermehrt ungeordneten Struktur unter dem Einfluss der Temperaturerhöhung schliessen. Die ungeordnete Struktur kann durch eine Vermischung der (2×1) und (2×2) Überstrukturen erklärt werden, die in einer Pseudo-(1×1)-Struktur resultiert. Gitterplatzverschiebungen der einzelnen Atome waren bis zu einer Tiefe von drei Einheitszellen signifikant.

Die *in-situ* kinetischen Untersuchungen des Wachstums von LSMO auf STO mittels gepulster Laser Abscheidung führten zur Formulierung eines neuartigen energieabhängigen Glättungsmechanismus für das Wachstum komplexer Metalloxidschichten. Unterhalb eines Bedeckungsgrades von 50% bezogen auf eine Monolage bewirkt der unmittelbare Einbau der einfallenden energiereichen Teilchen in die kleinflächigen Inseln ein Aufbrechen dieser zu Tochterinseln. Dieser Glättungsmechanismus unterbindet daher das Auftreten von grossflächigen 2D-Inseln und das Einsetzen von 3D-Wachstum. Die Inseln beginnen oberhalb von 50% Bedeckung zu verschmelzen, wobei ein Aufbrechen dadurch verhindert wird. Die Energie des ankommenden Teilchenstrahls wird anstelle dessen für gesteigerte Oberflächendiffusion genutzt.

Die Strukturen von sechs dünnen LSMO Schichten wurden bestimmt. Die Schichten wurden mittels gepulster Laser Abscheidung Monolage-für-Monolage gewachsen und wiesen Dikken zwischen 1 und 9 Einheitszellen auf. Als Ausgangspunkt für die Strukturbestimmung wurden die Resultate der vorangegangenen COBRA Phasenbestimmung benutzt. Folgende vier Resultate konnten gewonnen werden. Erstens wurde eine Ausdehnung der Gitterkonstante normal zur Substrat-Schicht-Oberfläche beobachtet. Zweitens ist der Übergang von Substrat in den Film nicht abrupt, sondern kontinuierlich über etwa drei Einheitszellen. Drittens segregiert Sr in die oberste Monolage der Schicht. Die Verwendung der Besetzungszahlen liess eine Bestimmung der Segregationsenthalpie von Sr auf -15 kJ/Mol zu. Viertens konnte die elektronische Bandbreite W zur Erklärung des Auftretens von Magnetwiderstand für Schichtdicken von 9 und mehr Monolagen verwendet werden. Messungen des spezifischen Widerstandes an der Schicht von 9 Monolagen Dicke bestätigen das Auftreten von Magnetwiderstand und die Anwesenheit einer toten Schicht mit mehrheitlich isolierendem Verhalten.

Copyright of Papers I, II, IV, and V is held by the American Physical Society. The reprints are with permission of the publisher as follows.

Paper I: Reprinted with permission from R. Herger, P.R. Willmott, O. Bunk, C.M. Schlepütz, B.D. Patterson, and B. Delley, *Surface of strontium titanate*, Phys. Rev. Lett. **98**, 076102 (2007). Copyright (2007) by the American Physical Society.

Paper II: Reprinted with permission from R. Herger, P.R. Willmott, O. Bunk, C.M. Schlepütz, B.D. Patterson, B. Delley, V.L. Shneerson, P.F. Lyman, and D.K. Saldin, *Surface structure of $SrTiO_3(001)$*, Phys. Rev. B. **76**, 195435 (2007). Copyright (2007) by the American Physical Society.

Paper IV: Reprinted with permission from P.R. Willmott, R. Herger, C.M. Schlepütz, D. Martoccia, and B.D. Patterson, *Energetic surface smoothing of complex metal-oxide thin films*, Phys. Rev. Lett. **96**, 176102 (2006). Copyright (2006) by the American Physical Society.

Paper V: Reprinted with permission from R. Herger, P.R. Willmott, C.M. Schlepütz, M. Björck, S.A. Pauli, D. Martoccia, B.D. Patterson, D. Kumah, R. Clarke, Y. Yacoby, and M. Döbeli, *Structure determination of monolayer-by-monolayer grown $La_{1-x}Sr_xMnO_3$ thin films and the onset of magnetoresistance*, Phys. Rev. B **77**, 084501 (2008). Copyright (2008) by the American Physical Society.

Springer and the original publisher Applied Physics A hold the copyright of

Paper III: Reprinted with kind permission from Springer Science and Business Media: P.R. Willmott, R. Herger, M.C. Falub, L. Patthey, M. Döbeli, C.V. Falub, M. Shi, and M. Schneider, *Pulsed laser deposition of atomically flat $La_{1-x}Sr_xMnO_3$ thin films using a novel target geometry*, Appl. Phys. A **79**, 1199 (2004).

Contents

Acknowledgments		vii
Abstract		ix
Zusammenfassung		xi
Contents		xv
List of Figures		xvii

1 Introduction — 1
    Bibliography 3

2 Transition metal oxides — 5
    2.1 Perovskites 6
    2.2 Crystallographic structure of perovskites 8
    2.3 Strontium titanate 10
    2.4 Lanthanum strontium manganate 11
    2.5 Application in thin films 16
    Bibliography 19

3 Theory — 25
    3.1 Diffraction 25
        3.1.1 Single crystal diffraction 25
        3.1.2 Crystal truncation rods 28
    3.2 Superstructure rods 30
    3.3 Direct methods 31

		3.3.1	PARADIGM	32
		3.3.2	COBRA	33
	Bibliography			37

4 Experiment — 39

	4.1	Pulsed laser deposition		39
		4.1.1	Setup	39
		4.1.2	Substrate preparation	40
		4.1.3	Growth	41
	4.2	Surface x-ray diffraction		42
		4.2.1	Diffractometer	42
		4.2.2	Detector	42
		4.2.3	Reciprocal space scans	44
	4.3	Atomic force microscopy		46
	4.4	X-ray photoemission spectroscopy		46
	4.5	Rutherford backscattering		47
	4.6	Resistivity		47
	Bibliography			49

5 Concluding remarks — 51

Paper I: Surface of strontium titanate — 55

Paper II: Surface structure of SrTiO$_3$(001) — 61

Paper III: Pulsed laser deposition of La$_{1-x}$Sr$_x$MnO$_3$ thin films — 81

Paper IV: Energetic surface smoothing of complex metal-oxide thin films — 87

Paper V: Structure determination of La$_{1-x}$Sr$_x$MnO$_3$ thin films — 93

Publications by R. Herger — 105

List of Figures

2.1	Super exchange and double exchange	7
2.2	Unit cells of $SrTiO_3$ and $La_{1-x}Sr_xMnO_3$	8
2.3	Electronic bandstructure diagrams of $SrTiO_3$, $La_{1-x}Sr_xMnO_3$, and Ni	9
2.4	Energy levels of the JT-active Mn ion	12
2.5	Schematic of double exchange of canted spins	13
2.6	Charge and orbital ordering: FM ($x = 1/3$) and AFM ($x = 1/2$)	15
2.7	Schematic of a FET on a p-type substrate	17
3.1	Ewald construction	27
3.2	Crystal truncation rods	28
3.3	Iterative phase recovery algorithm PARADIGM	32
3.4	COBRA real space origin	34
3.5	Graphical solution of Eqn. 3.21 of the COBRA approach	35
3.6	Flow diagram of the COBRA algorithm	36
4.1	*Ex-situ* and *in-situ* PLD chambers	40
4.2	$SrTiO_3$ etch process	41
4.3	Surface diffractometer	43
4.4	A typical Pilatus I image	44
4.5	Measurement principle of crystal truncation rods	45
4.6	A CTR cutting through the Ewald sphere	46
4.7	Four-point method	47

Chapter 1

Introduction

The advent of highly brilliant third generation synchrotron sources had an enormous impact on the research in the fields of both materials and life sciences. Biologists and chemists nowadays use synchrotron radiation routinely to achieve a better understanding of complicated structures like proteins and cells, while environmental issues are partly clarified by the incisive probe of synchrotron radiation. On the other hand, physicists and material scientists are motivated to study materials of challenging structural complexity and fascinating electronic properties. Moreover, the high brightness of synchrotron radiation, plus the well-shaped time structure of the electron bunches in the storage ring enable todays researchers to perform time-resolved measurements that one could not contemplate years ago.

The Swiss Light source is a state-of-the-art medium energy (2.4 GeV) third generation synchrotron source, located the the Paul Scherrer Institut, Villigen, Switzerland. It started its operation in 2001 and is, at the time of writing, running 12 beamlines (BLs). The work described in this thesis was mainly performed at the Surface Diffraction station of the Materials Science BL [1]. The high flux of this wiggler BL, a fast single photon-counting 2-dimensional (2D) x-ray pixel detector [2], and a pulsed laser deposition chamber that can be mounted on a large $(2+3)$ circle diffractometer [3] are ideally combined to study *in-situ* both surface and interface structures, as well as the kinetics and evolution of thin film growth of novel materials.

The work presented here is mainly based on surface x-ray diffraction (SXRD), where the data was collected using the 2D pixel detector – a novelty to this community [4], and an established adaptation to pulsed laser deposition where the ablation plasma is crossed by a reactive gas pulse [5], in the present case an oxidant to enhance the oxygen content of the deposited complex metal oxide.

The thesis presents structure determinations and a kinetic growth study of two materials, strontium titanate $SrTiO_3$ (STO, a prototypical substrate material) and lanthanum strontium manganate $La_{1-x}Sr_xMnO_3$ (LSMO, grown as thin films [6]). STO and LSMO are perovskites that can generally be described by the chemical sum formula ABO_3, where A is an alkaline or rare earth and B is usually a transition metal. The thesis is structured as follows.

In Chapter 2, a general introduction to transition metal oxides (TMOs) and perovskites is presented. In particular, the important physical properties and possible technological applications are highlighted and reviewed in the literature.

A theoretical background of SXRD as well as an overview of the phase retrieval techniques PARADIGM and COBRA is given in Chapter 3, as they were of importance for the structural determinations in this work.

Experimental work is described in Chapter 4 in order to detail and complement the condensed information that can be found in the subsequently presented reprints of the literature.

Concluding remarks are presented in Chapter 5.

Paper I describes the surface structure determination of TiO_2-terminated STO(001) under two different conditions: one at room temperature in vacuum and the other hot, under typical conditions for thin film growth in an oxygen atmosphere [7].

The details of the surface structure determination of STO were subsequently described in a more comprehensive article [8]. This can be found in Paper II. R.H. was the main responsible for preparatory, experimental, analysis and refinement work of the STO publications, assisted by B.D.'s density functional theory calculations and the PARADIGM work in a collaboration with V.L.S., P.F.L. and D.K.S. of the University of Wisconsin-Milwaukee.

In Paper III, the experimental details of pulsed laser deposition of LSMO thin films is described, detailing typical growth conditions and *ex-situ* characterization of the films [6].

Paper IV describes a kinetic study of the growth of LSMO, where a novel energetic surface smoothing mechanism for the growth of TMOs by pulsed laser deposition was proposed [9]. R.H. was mainly responsible for substrate processing, film growth, preparation and realization of the beamtime, whereas P.R.W. did the successive analysis work.

In Paper V the structure determinations of monolayer-by-monolayer grown LSMO thin films were used to explain the onset of magnetoresistance [10]. R.H. was responsible for the experiments, analysis and interpretation. The structure determination using COBRA was done at the Randall Laboratory of Physics at the University of Michigan (D.K. and R.C.) and at the Racah Institute of Physics at the Hebrew University in Jerusalem (Y.Y.).

Bibliography

[1] B. D. Patterson, R. Abela, H. Auderset, Q. Chen, F. Fauth, F. Gozzo, G. Ingold, H. Kühne, M. Lange, D. Maden, D. Meister, P. Pattison, T. Schmidt, B. Schmitt, C. Schulze-Briese, M. Shi, M. Stampanoni, and P. R. Willmott, *The Materials Science Beamline at the Swiss Light Source: design and realization*, Nucl. Instrum. Methods A **540**, 42 (2005).

[2] C. Brönnimann, S. Florin, M. Lindner, B. Schmitt, and C. Schulze-Briese, *Synchrotron beam test with a photon-counting pixel detector*, J. Synchr. Rad. **7**, 301 (2000).

[3] P. R. Willmott, C. M. Schlepütz, B. D. Patterson, R. Herger, M. Lange, D. Meister, D. Maden, C. Brönnimann, E. F. Eikenberry, G. Hülsen, and A. Al-Adwan, *In situ studies of complex PLD-grown films using hard x-ray surface diffraction*, Appl. Surf. Sci. **247**, 188 (2005).

[4] C. M. Schlepütz, R. Herger, P. R. Willmott, B. D. Patterson, O. Bunk, C. Brönnimann, B. Henrich, G. Hülsen, and E. F. Eikenberry, *Improved data acquisition in grazing-incidence x-ray scattering experiments using a pixel detector*, Acta Crystallogr. Sect. A **61**, 418 (2005).

[5] P. R. Willmott and J. R. Huber, *Pulsed laser vaporization and deposition*, Rev. Mod. Phys. **72**, 315 (2000).

[6] P. R. Willmott, R. Herger, M. C. Falub, L. Patthey, M. Döbeli, C. V. Falub, M. Shi, and M. Schneider, *Pulsed laser deposition of atomically flat $La_{1-x}Sr_xMnO_3$ thin films using a novel target geometry*, Appl. Phys. A **79**, 1199 (2004).

[7] R. Herger, P. R. Willmott, O. Bunk, C. M. Schlepütz, B. D. Patterson, and B. Delley, *Surface of strontium titanate*, Phys. Rev. Lett. **98**, 076102 (2007).

[8] R. Herger, P. R. Willmott, O. Bunk, C. M. Schlepütz, B. D. Patterson, B. Delley, V. L. Shneerson, P. F. Lyman, and D. K. Saldin, *Surface structure of SrTiO$_3$(001)*, Phys. Rev. B **76**, 195435 (2007).

[9] P. R. Willmott, R. Herger, C. M. Schlepütz, D. Martoccia, and B. D. Patterson, *Energetic surface smoothing of complex metal-oxide thin films*, Phys. Rev. Lett. **96**, 176102 (2006).

[10] R. Herger, P. R. Willmott, C. M. Schlepütz, M. Björck, S. A. Pauli, D. Martoccia, B. D. Patterson, D. Kumah, R. Clarke, Y. Yacoby, and M. Döbeli, *Structure determination of monolayer-by-monolayer grown La$_{1-x}$Sr$_x$MnO$_3$ thin films and the onset of magnetoresistance*, Phys. Rev. B **77**, 085401 (2008).

Chapter 2

Transition metal oxides

Transition metal oxides (TMOs) often exhibit strong interactions between the electrons due to Coulomb repulsion. They show unusual electronic and magnetic properties due to the strong coupling of the electrons. They typically have partially filled d or f shells. Strongly correlated electron systems (SCES) have electrons that neither can be sufficiently described by a one electron approximation where the electron is free to move in the mean field of the other electrons of the atom, nor completely ionic, i.e., bound to the nucleus. Instead, the electronic structure of SCES have to be described with models that take the mobility of the electrons into account, as well as the correlation between them. The Hubbard model (presented below) is such a model. SCES are characterized by the complex, often simultaneously active interplay of their degrees of freedom: spin, charge, lattice and orbital interactions. The strong electron correlation is of importance to understand the metal-insulator transitions in TMOs and causes fascinating physical phenomena, such as high-temperature superconductivity, colossal magnetoresistance, Mott insulation, or ferroelectric behavior, to name only a few examples [1–4].

The quantum mechanical wavefunction used to describe an electron takes different shapes when bound to an atomic nucleus by Coulomb force. Let us consider a transition metal atom in an octahedral crystal field, i.e., surrounded by six adjacent oxygen atoms in the case of a perovskite. The spatial orientation of the wavefunction produces different responses to a perturbation from the neighboring oxygens. In other words, the electron correlation will be different. depending on whether the wavefunction points directly at the neighboring oxygens or exactly in between them. To gain an understanding of the underlying physics in TMOs, it is therefore important to know the exact atomic positions.

So far, we have only described an isolated transition metal ion. But in crystalline solids, periodic arrays of the ions lead to magnetic interactions (interacting spins) or band formation and electric conduction (interacting electrons in partially filled bands). Insulators are normally characterized by their having completely filled or empty bands. An exception is the *Mott insulator*, which band theory predicts to be a conductor, but in fact is insulating, due to the strong electron correlation: the $3d$ electrons are well localized (and not free to move) because their kinetic energy to hop to a neighboring site is too small to overcome the strong Coulomb repulsion energy of the electrons. This competition is described in the Hubbard model [5], which, in the simplest case, has a "one-band" Hamiltonian (one orbital per site) of the form:

$$H = -t \sum_{<i,j>,\sigma} c^\dagger_{i,\sigma} c_{j,\sigma} + U \sum_i n_{i\alpha} n_{i\beta}, \qquad (2.1)$$

where $<i,j>$ represents nearest neighbor-sites on the lattice. $c^\dagger_{i,\sigma}$ and $c_{j,\sigma}$ create and annihilate an electron with a z component of spin σ at site i or j, respectively, and $n_{i\sigma} = c^\dagger_{i,\sigma} c_{i,\sigma}$ gives the number of spin σ electrons at site i. The electron transfer integral t is termed hopping term and the Coulomb repulsion U called the interaction term. A Mott insulator thus has a large U/t value.

2.1 Perovskites

A class of particularly important TMOs has the general formula ABO_3. They have mostly the perovskite structure and are named after the mineral "perovskite" $CaTiO_3$ for which this atomic arrangement was first found. In perovskites, the A-site is usually occupied by an alkaline-earth or rare-earth metal, whereas the B-site normally contains a transition metal, a rare-earth or a group III metal. The B-site is surrounded by six oxygen forming an octahedron. It is the BO_6-octahedron that is mostly responsible for the intriguing physical properties of perovskites. Let us exemplify this by inspection of the transition metal group in the periodic table of the elements. Some perovskite-based transition metal compounds with a B-site consisting of Cu, Ni and Fe, Co, and Mn exhibit high-temperature superconductivity [1], cooperative charge and spin ordering [6], enhanced thermopower [7], and colossal magnetoresistance [3], respectively.

The crystal field of the six neighboring O-atoms cause the five initially degenerate d orbitals to split into an energetically lower, 3-fold degenerate t_{2g} (d_{xy}, d_{xz}, and d_{yz}) set and in a higher 2-fold degenerate e_g ($d_{x^2-y^2}$ and $d_{3z^2-r^2}$ [1]) state. The splitting energy is named 10Dq or Δ (used in this work) and is typically of the order of a few eV, i.e., comparable to the energy

---
[1] The notation for the $d_{3z^2-r^2} = d_{2z^2-x^2-y^2}$ is often simplified to d_{z^2}.

2.1. PEROVSKITES

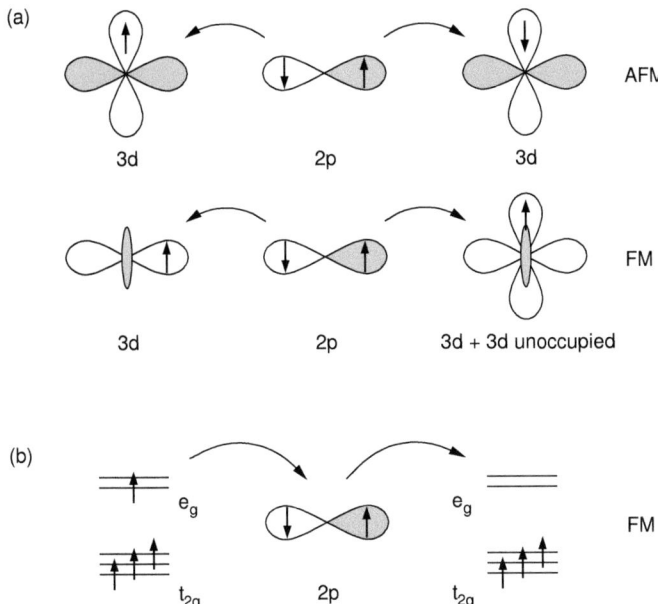

Figure 2.1: Schematic representation of super exchange (a) and double exchange (b). Super exchange: Chemical bonding interactions result in AFM-coupled spins (top) for neighboring $3d$ orbitals. The exchange interaction is mediated through the electron spins in the ligand $2p$ orbital. For FM-coupling between the spins (bottom), a filled orbital (here either $3d_{xz}$ or $3d_{yz}$) has to be orthogonal to the unfilled orbital (here $3d_{3z^2-r^2}$) and has the same electron spin directions (Hund's rule coupling). The double exchange mechanism transfers charge: The e_g electron hops to the ligand $2p$ *simultaneously* as one $2p$ electron hops to unfilled e_g band of the neighboring metal ion.

of a chemical bond. Depending on the band-filling the octahedral crystal field can undergo a Jahn-Teller distortion, as will be detailed in Section 2.2.

Here, two important exchange interactions for understanding the physics of perovskites are introduced: super exchange (SE) [8] and double exchange (DE) [9]. In both cases, neighboring perovskite B-sites interact via the connecting oxygen $2p$ orbital. The main difference between the two mechanisms is that in one case, a virtual electron transfer mediates the magnetic [antiferromagnetic (AFM) or ferromagnetic (FM)] interaction (SE), but in the other case, charges

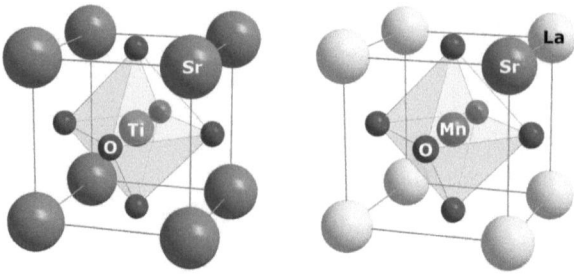

Figure 2.2: Atomic arrangement of the perovskite structure unit cell for SrTiO$_3$ (STO) and La$_{1-x}$Sr$_x$MnO$_3$ (LSMO). Sr or La/Sr, and Ti or Mn occupy the A- and B-sites, respectively. (Pseudo-) Cubic lattice constants are $a_{\text{STO}} = 3.905$ and $a_{\text{LSMO}} = 3.875$ Å (at $x = 0.35$).

are effectively transferred between two metal ions (DE) and thus the long-range FM order is mediated. Both exchange mechanisms are schematically shown in Fig. 2.1. Note that DE implies a mixed valent system, whereas SE interactions can also occur for metal ions with equal valence states.

2.2 Crystallographic structure of perovskites

Two unit cells of perovskites, SrTiO$_3$ and La$_{1-x}$Sr$_x$MnO$_3$, are shown in Fig. 2.2. The coordination number of perovskites are 12 (8) and 6 (5) for A- and B-sites in bulk (surface) positions, respectively. The typical (cubic) perovskite lattice constant is approximately $a = 3.9$ Å. Atoms of the idealized cubic unit cell occupy the typical positions at $(0,0,0)$ (A-site), $(1/2,1/2,1/2)$ (B-site), and $(0,1/2,1/2)$, $(1/2,0,1/2)$, and $(1/2,1/2,0)$ (the three O-sites). The composition of such a perovskite is often visualized as stacks of alternating AO- and BO_2-layers.

The crystallographic structure of ABO_3 perovskites is governed by the ionic radii of the constituent elements [10]. The tolerance factor Γ is defined as

$$\Gamma = \frac{r_A + r_O}{\sqrt{2}(r_B + r_O)} \quad (2.2)$$

where r_A, r_B and r_O are the ionic radii of A, B and O, respectively. Stable perovskite structures have tolerance factors of approximately $0.75 < \Gamma < 1.05$. Cubic structures are observed for $0.89 \leq \Gamma \leq 1.00$, orthorhombic for $\Gamma < 0.89$ and rhombohedral for $\Gamma > 1.00$ [11, 12]. Structural

2.3. STRONTIUM TITANATE

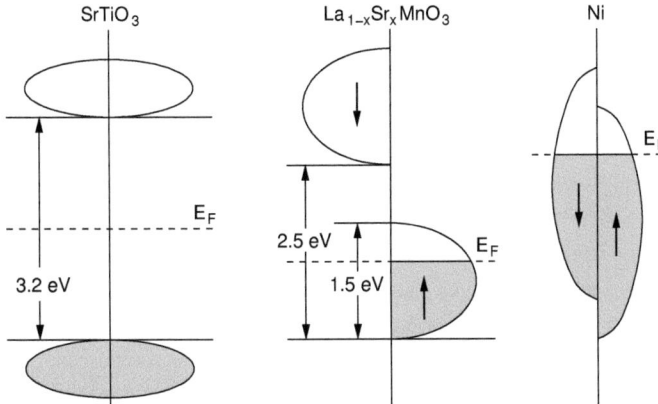

Figure 2.3: Schematic of the electronic bandstructures of SrTiO$_3$ (left), La$_{1-x}$Sr$_x$MnO$_3$ (middle) and, for comparison, Ni (right). Dark bands are filled, white bands empty, E_F denotes the Fermi energy, and the arrows indicate the electron spins. STO is a band gap insulator. LSMO is a half-metallic ferromagnet [14] where one electron spin (here up) shows metallic and the other (down) insulating behavior in coexistence. Ni, on the other hand, is a conventional ferromagnetic metal with simultaneously occupied up and down spin states.

distortions lead to a rotation of the BO_6 octahedron. This can have important consequences on the electronic and magnetic properties, as we will see later.

The degenerate t_{2g} and e_g states in an octahedral crystal field can be distorted to reduce the energy eigenvalues of certain d orbitals of a TMO. This is the so-called Jahn-Teller (JT) distortion [13]. It is most pronounced in octahedral complexes when the system contains odd numbers of electrons in the e_g state, e.g., d^9, low-spin (= maximum of paired electrons) d^7, and high-spin (= unpaired electrons) d^4. In these cases, the e_g orbitals point directly towards a ligand ion, and hence, a structural distortion can lead to a large energetic stabilization. For d^1 and d^2 transition metals, the JT distortion can also occur, but since the t_{2g} orbitals are in between the ligand atoms, the effect is much less pronounced.

2.3 Strontium titanate

Strontium titanate (SrTiO$_3$, STO) is a band-gap insulator with a gap energy of 3.2 eV [15] (see Fig. 2.2 for the unit cell and Fig. 2.3 for a schematic of the bandstructure). Single-crystals are grown using the Verneuil-method [16].

STO has a high refractive index of $n = 2.41$ (at 589 nm), nearly identical to that of diamond, and a density of 5.117 g cm^{-3}. It has a very high dielectric constant $\varepsilon_r = 300$, making it useful for the thin film fabrication of high-voltage capacitors.

Bulk STO undergoes a structural phase transition from its cubic phase to a tetragonal phase at a temperature of $T = 105$ K (space group $Pm\bar{3}m$ to $I4/mcm$). Experimentally, this phase transition was determined by electron spin resonance and Raman spectroscopy [17, 18]. The transition involves a rotation of the TiO$_6$ octahedron with the rotation angle as the order parameter [19]. This structural transition is accompanied by a ferroelectric relaxation [20, 21]. Below 0.3 K, STO becomes superconducting [22].

STO is well-known as the paradigmatic substrate material for the growth of perovskite-based thin films showing effects such as those mentioned at the beginning of this chapter. This is mainly due to its good lattice match, the high crystallinity[2], the flat surface, and the high $p4mm$ surface symmetry.

Polished STO surfaces naturally exhibit a mixed termination of TiO$_2$ and SrO with SrO percentages ranging from 5 to 25% [23, 24]. This is, however, not suitable if one intends to grow materials with distinct properties in a controlled way. Therefore, standard processing conditions for the STO surface have been established in order to achieve single TiO$_2$-terminated STO surfaces with terrace 4 Å step heights [25–28].

The surfaces of unprocessed, as well as chemically and/or thermally treated STO have attracted much attention in the literature due to the rich variety of different surface terminations concerning reconstructions and relaxations; see Ref. [29] for an overview. It seems that the surface of bulk STO is very sensitive to the preparation and processing conditions, leading to various reactions of the surface to these perturbations. Several concepts have been used to explain the experimental observations: lateral displacements including rumpling and buckling of the surface atoms [20, 21, 30], oxygen deficiency [31–33], an oxygen overlayer [34], a Sr-adatom model [35], surfaces containing nano-islands consisting either of SrO or TiO [36, 37], surfaces containing a double TiO$_2$-layer [28, 38, 39], or (partially) SrO-terminated surfaces

---

[2] "High crystallinity" in comparison to e.g., metals, although it is noted that Czochralski-grown perovskites such as NdGaO$_3$ or DyScO$_3$ tend to be of superior crystal quality.

[36, 40, 41]. Importantly, these terminations exist in distinct temperature and pressure regimes under either oxidizing or reducing conditions. In thermodynamical equilibrium, however, the surface should be SrO-terminated, as has recently been suggested on the basis of theoretical arguments [42]. The observation of the formation of SrO-islands for high annealing temperatures or long annealing times partly supports this [36, 41]. Depending on the application, the presence of a kinetically hindered TiO_2-terminated surface is preferred over the SrO-termination of the thermodynamical equilibrium. A good illustrative example is the presence of a quasi two-dimensional electron gas at the interface of STO with $LaAlO_3$, both insulators: The electron gas is only present in the case of TiO_2-terminated STO, whereas for SrO-termination, no such effect is observed [43].

The surface of strontium titanate, which at first glance might seem to be simple, still challenges science experimentally, methodically and theoretically, even if the bulk properties have been known for over 50 years. In particular, it is the heterointerface to other materials that has recently stimulated research, as we will see in the next section.

2.4 Lanthanum strontium manganate

Ferromagnetic (FM) manganites were first investigated by Jonker and van Santen in 1950, but have attracted renewed interest in recent years since the discovery that they exhibit colossal magnetoresistance (CMR), i.e., the very-large magnetic-field driven change in resistivity [3, 44]. Doped manganites with the perovskite structure and chemical composition $Re_{1-x}Ae_xMnO_3$, where Re is a rare-earth and Ae a divalent alkaline-earth, show rich phase diagrams, due to the complex interplay of charge, spin, lattice and orbital degrees of freedom [45, 46]. Their interesting physical properties have not only triggered renewed scientific interest in these compounds but also show potential for many technological applications, such as spin electronics or magnetic sensors. Thin films are best suited for these demands.

Bulk $La_{1-x}Sr_xMnO_3$ (LSMO) at an optimal doping[3] of $x = 1/3$ shows the transition from a paramagnetic insulator to a ferromagnetic metal at $T_C = 370$ K [46]. The Mn site has a mixed valence state of x Mn^{4+} (holes) and $(1-x)$ Mn^{3+}, leading to degenerate high-spin t_{2g}^3 and $t_{2g}^3 e_g^1$ states of the MnO_6 octahedra, respectively, due to the large Hund's rule coupling energy[4]

---

[3] A simple model using the mean field approximation leads to the following relation between the Curie temperature T_C and the doping level x: $T_C \propto 2x(1-x) - x^2$, which is maximum for $x = 1/3$. This crude model agrees with the fact that FM phases are generally found around $x = 1/3$ in manganites. For details, see Ref. [12].

[4] To account for Hund's rule coupling, the Hamiltonian of Eqn. 2.1 has to be rewritten as shown in Ref. [47].

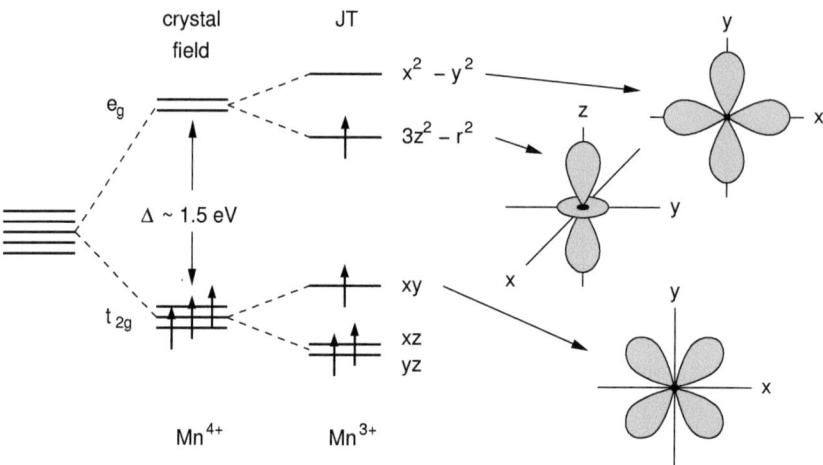

Figure 2.4: Energy levels of the JT-active Mn ion. The Mn^{4+} is shown in the octahedral crystal field, filling the t_{2g} states (total spin $S = 3/2$). The Mn^{3+} has an additional unpaired electron ($S = 2$, high-spin state) and undergoes a JT distortion with axial elongation.

of $J_H \approx 2.5$ eV compared to the crystal field Δ (see Fig. 2.4). The electrons can hop between adjacent Mn ions, as described by Zener's double-exchange mechanism [see Fig 2.1(b)] [9] and thus mediate the long-range FM ground state of the metallic conductor. Additionally, as the electronic ground states of the Mn^{3+} sites are degenerate, a Jahn-Teller distortion breaks the octahedral symmetry and lowers the energy (see Fig. 2.4) [48].

At a fixed hole density x, the properties of LSMO and manganites are affected by distortions of the ideal cubic geometry, qualitatively expressed by the tolerance factor (Eqn. 2.2). The transport properties depend on the overlap of the Mn sites with the O $2p$ orbitals, which in turn is determined by the Mn–O–Mn angle ϕ. For $\phi < 180°$, a reduced electron hopping amplitude results, which is proportional to $\cos\phi$ [49]. In the case of LSMO, the tolerance factor $\Gamma = 0.98$ (for $x = 0.35$), $\phi < 180°$, and the tendency towards charge localization increases. Furthermore, in the Hubbard picture (Eqn. 2.1), the electron hopping term t not only depends on the bond angle ϕ, but also on the length of the Mn–O bond as $1/(d_{\text{Mn–O}})^\alpha$, where $\alpha > 1$ [50]. We

---

The term account for Hund's rule coupling becomes to $-J_H \sum_{i,\alpha,\beta} \mathbf{s}_i \cdot \mathbf{S}_i$, where $\mathbf{s}_i = c^\dagger_{i,\alpha} \sigma_{\alpha,\beta} c_{i,\beta}$ creates an electron at site i with spin α. $\sigma_{\alpha,\beta}$ are the Pauli spin matrices. $\mathbf{S}_i$ is the total spin at site i.

2.4. LANTHANUM STRONTIUM MANGANATE

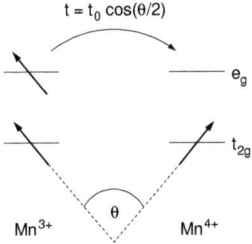

Figure 2.5: Schematic view of the DE mechanism of canted spins illustrating the electron transfer of $t = t_0 \cos(\theta/2)$. The oxygen is omitted for simplicity.

therefore expect changes in T_C when ϕ and/or $d_{\text{Mn-O}}$ change.

The magnetic properties of the manganites are affected by the exchange interactions between the Mn ion spins (Fig. 2.1). In general, the Mn^{4+}–O–Mn^{4+} interaction is antiferromagnetic (AFM), whereas for Mn^{3+}–O–Mn^{3+} it may be FM or AFM [51]. The DE interaction of Mn^{3+}–O–Mn^{4+} is FM in nature. Anderson and Hasegawa showed [52] that in contrast to the angular dependence of the usual exchange interaction (proportional to $\cos\theta$), the probability of the e_g electron to hop from the Mn^{3+} to the Mn^{4+} is $t = t_0 \cos(\theta/2)$, with θ defined as the angle between neighboring spins, and under the assumption of a large J_H (Fig. 2.5). Parallel spin configuration ($\theta = 0$) thus maximizes t, whereas for an AFM spin arrangement ($\theta = \pi$), no DE process takes place. In summary, the DE energy parameter for manganites is [53]:

$$J \propto x(1-x)\cos\phi\cos(\theta/2). \qquad (2.3)$$

These different angular dependencies in combination with the competition between DE-FM and SE-AFM states lead to the rich magnetic phase diagrams of manganites [45, 46].

The DE exchange process is the basic mechanism for the observed electrical conduction in the FM state at low temperatures in the manganites. It involves a charge transfer that leads to a delocalization of the e_g states for a certain range of doping centered around $x = 1/3$. Figure 2.3(center) shows the schematic bandstructure of the FM state of LSMO. The filling of the spin-up band of LSMO represents the $(1-x)$ e_g electrons that can electrically conduct. The bandwidth of about 1.5 eV is smaller than the Hund's rule coupling energy J_H, thus leaving the upper spin-down band empty. These bandstructure features are termed a half-metallic ferromagnet, which was experimentally observed by Park *et al.* for LSMO [14]. The conduction band of such a half-metallic ferromagnet is fully spin-polarized, explaining the interest in these compounds for application as spin electronics. For comparison, the bandstructure of a $3d$ ferromagnetic metal such as Ni or Fe is shown. Here, both bands are simultaneously occupied with up and down spins.

A typical resistivity measurement of a thin film of LSMO of 130 nm thickness without applied magnetic field is given in Fig. 4 of Paper III. Crossing the Curie temperature T_C, the resistivity changes from paramagnetic semiconducting or insulating behavior to a low temperature FM phase. The application of a magnetic field changes T_C to higher temperatures and lower electrical resistances R. Defining the magnetoresistance ratio as $\Delta R/R_B = (R_B - R_0)/R_B$, where the subscripts indicate resistances measured with applied field or without, this ratio becomes negative for materials exhibiting giant magnetoresistance (typically -5 to -50%) or CMR (minus several orders of magnitude) [3], in contrast to ordinary permalloys ($\Delta R/R_B \approx +5\%$). The negative sign and the (usually) isotropic response to the direction of the applied field are characteristic for these materials showing extraordinary magnetoresistive effects.

Since the early studies of the manganites in the 1950s and the discovery of CMR, the understanding of the underlying physics has changed dramatically. First attempts were based on the DE mechanism and the strong Hund coupling to qualitatively explain the increase in conductivity upon the polarization of the spins [52]. Work by de Gennes using mean-field approximations suggested that the interpolation between the AFM phase for $x = 0$ and the FM state at finite, sufficiently low level doping x (i.e., described by the DE mechanism) occurs through a canted state of the spins, as shown in Fig. 2.5 [54]. This idea proved to be insufficient to quantitatively explain the CMR effect. One of the reasons for incompleteness of this early theoretical idea was its inability to account for the electron-phonon interaction. Millis *et al.* [48] were the first to claim that any quantitative explanation needs a more elaborate description, where a large JT effect produces a strong electron-phonon interaction that persists in the FM phase. Experimental evidence for the electron-phonon interaction was given for instance from the large oxygen isotope effect observed for T_C [55]. Recent theoretical work has shown that the spin-canted state proposed by de Gennes is not realized in the model of manganites. Instead, phase separation between the FM and the AFM states occur, where FM metallic and AFM charge and orbital ordered domains are spatially separated, i.e., manganites are *intrinsically inhomogeneous*. Experimentally, phase separation was observed using direct space techniques such as scanning tunneling spectroscopy or electron microscopy [56–58]. The phase separation scenario has been reviewed extensively [49, 59].

One of the most comprehensive theoretical understanding of CMR, both qualitatively and quantitatively, was recently given by Şen *et al.* [63]. The combination of the DE mechanism,

2.4. LANTHANUM STRONTIUM MANGANATE

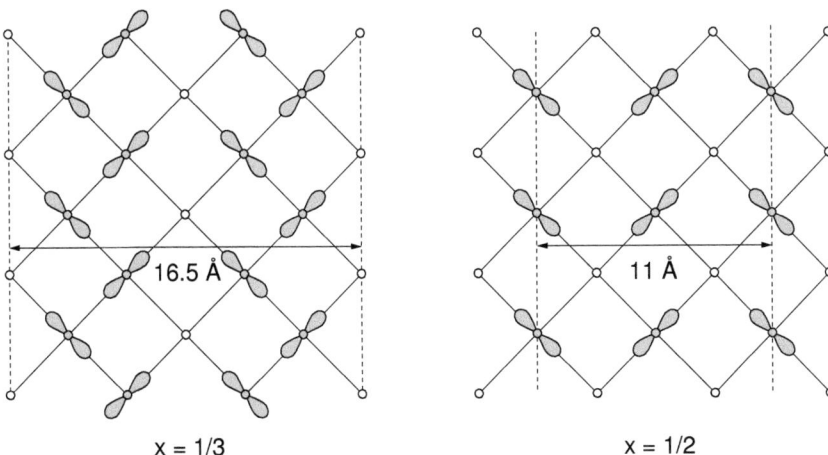

Figure 2.6: Charge and orbital ordering configurations for $La_{1-x}Sr_xMnO_3$ viewed along (001). A FM metallic state was proposed to form stripes for optimal doping $x = 1/3$ [60], whereas an AFM metallic state (A-type spins) in LSMO occurs for $x = 1/2$ [61]. Open circles represent Mn^{4+}, and the lobes show the orbital ordering of Mn^{3+} given in dark closed circles. The spins (not shown in the schematic) arrange as follows. For the FM state, all spins of the top and all lower layers point into the same direction, whereas for the A-type AFM arrangement [62], the spins of the top layer point in one direction, but the spins of the next deeper layer point to the opposite direction, and so on.

electron-phonon interaction and AFM super exchange coupling J_{AFM} can explain the CMR effect as follows: Imagine, in a phase separation scenario, coexisting FM metallic domains and charge-ordered AFM states (see Fig. 2.6). Above T_C, nanoscale short-range ordered regions form, which exhibit the same charge and spin pattern as the low temperature insulating charge-ordered AFM state. Reducing the temperature below T_C will reduce this AFM-type short-distance charge-order, but the state does not disappear [63]. The FM order of the spins combined with the reduced correlation of the charges (i.e., they are more mobile) causes the drop in the electrical resistance of manganites. To obtain CMR effects, these charge-ordered clusters can be as small as a few unit cells. Nevertheless, the fact that the holes are not randomly distributed over short distances (i.e., it is not a gas of heavy polarons), but are indeed ordered, means that *correlated polarons* are at the origin of CMR.

2.5 Application in thin films

The progress made in synthesis, characterization and theory of TMOs is converging and leads to two main fields of opportunities for the application of thin TMO films in devices: the atomic-scale control of physical effects, and the combination of different functional oxides in direct contact that may give rise to novel physical effects and applications. Some of the interesting possibilities using thin LSMO films in technical devices will be presented in this section.

First, the fundamental difference between material in the bulk form and applied as thin films has to be recalled. LSMO, for instance, shows different electric transport properties for thin films and bulk material, the latter resembling granular ferromagnets. Moreover, surface and interface effects can set a lower limit to downsizing devices that exploit bulk effects. On the other hand, thin films allows one to study finite thickness effects, biaxial strain, interface and proximity effects, to name only a few examples, that have direct consequences on the physical behavior with sometimes surprising results. A sound fundamental understanding of the physics happening at the length scale of inter-atomic distances of a few Ångströms is necessary and could lead to devices with improved versatility and/or performance.

A possible application of LSMO in magnetoresistive read heads, for example, or high-sensitivity elements of non-volatile, magnetic random access memories in computer technology demands a further increase in the bit density per unit area, and thus the ability to reliably replicate patterns of the size of about 50 to 100 nm or lower. Here, standard photolithographic processing with typical pattern sizes of 1 micron are no longer suited, and the development of nanolithographic techniques such as scanning electron beam lithography is an active field of applied research [12].

One of the most promising fields of application for LSMO thin films is their use in tunnel junctions, because of their high degree of spin polarization and the ability of spin-dependent tunneling across a thin insulating layer. This makes them attractive for true on-off operations in computer memories as well as spin valves in magnetic sensors or read heads of hard disks. The difference in resistance of a trilayer consisting of LSMO/STO/LSMO was found to induce a change in magnetoresistance by tunneling as high as 83% at 4.2 K between parallel and antiparallel configurations of the magnetic moments in LSMO with an applied field of the order of 0.01 T [64]. The drawback is, however, the drastic reduction of the tunneling magnetoresistive effect when increasing the temperature (basically zero above 220 K), which in turn destroys the possibility of applying such a device under ambient conditions. On the other hand, the Fe/MgO/Fe junction was found to have a tunnel magnetoresistance ratio up to 180%

2.5. APPLICATION IN THIN FILMS

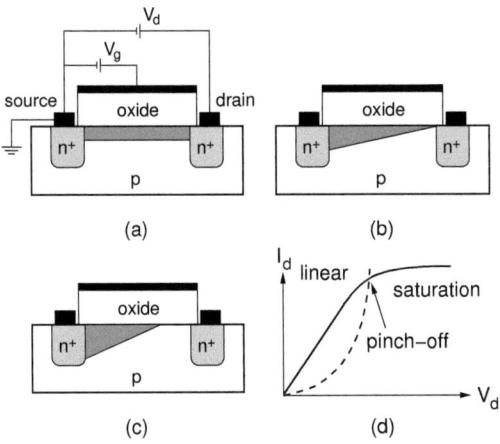

Figure 2.7: Schematic of an oxide-based FET on a p-type (hole-doped) substrate, with the conduction channel being an inversion layer of electrons. The source and drain contacts are made from heavily doped n-type (electron-doped, the superscripted + indicates high doping level) terminals that ensure ohmic contact. The shading indicates the density of electrons. The three operation states are: (a) linear, $V_d \ll (V_g - V_t)$, (b) near pinch-off, $V_d = (V_g - V_t)$, and (c) saturated, $V_d > (V_g - V_t)$, where V_d, V_g, and V_t denote the drain, gate, and threshold voltage of the FET, respectively. For $V_d > (V_g - V_t)$, i.e., above the pinch-off, a portion of the channel is turned off and the current-voltage characteristics are no longer linear. The FET is turned off when $V_g < V_t$. (d) Solid-line IV-curve of a FET showing the linear region, the pinch-off point and saturation for a particular V_g. The dashed-line indicates currents where $V_d = (V_g - V_t)$.

at room temperature [65].

The application of the CMR effect in thin films as magnetic sensors at room temperature shows a much smaller dependence between resistance and the applied magnetic field[5] compared to conventional permalloy films. However, such devices can operate in a wide range of magnetic fields, and, even more importantly, the CMR effect, which is microscopic in nature, occurs on length scales of several unit cell sizes of LSMO (see Paper V) and does not involve large scale entities such as magnetic domains and walls.

A very active and growing area of research is engaged in multicomponent systems of lay-

---
[5] Remember that for the measurements of CMR typically fields of several Teslas are applied.

ered oxides with different functions. Typically, a ferroelectric is combined with one or more other functional oxides, such as the high-temperature superconductors or ferromagnetic manganites, to a bilayer or multilayers. The combination of physical effects led to term these systems as "multiferroic". These materials use the fact that the electrical polarization and the large electric fields produced by the spatial displacement of the atoms in the unit cell of the ferroelectric in direct contact with the ferromagnet can modulate the magnetoresistance of the manganite film on a quasi atomic level. A nice introduction to ferroelectricity in thin films and heterostructures can be found in Ref. [66].

The direct contact of ferroelectric materials with another oxide layer reduces the screening lengths of the charge carriers from hundreds of Ångströms in conventional semiconductor devices to one to three unit cells of the interface for oxide field effect transistors (FET) [67]. The principle of a FET is shown in Fig. 2.7.

Another possibility for the ferroelectric to influence the adjacent functional oxide is the inverse piezoelectric effect. An electric voltage applied the ferroelectric layer causes the piezoelectric strain to be transferred to the second layer [68].

The driving force for making such devices possible is the advances in thin film deposition techniques such as molecular beam epitaxy, sputtering, pulsed laser deposition or chemical vapor deposition. One can now routinely epitaxially grow single-crystalline perovskite-based oxide thin films of a few unit cell thickness within an accuracy of much less than a lattice constant and atomically flat over hundreds of square microns. This not only enables one to combine different oxides for the search of novel interesting physics, but also opens the opportunity to control effects on the length scale of a unit cell of a crystal or less.

Bibliography

[1] J. G. Bednorz and K. A. Müller, *Possible high-T_c superconductivity in the Ba-La-Cu-O system*, Z. Phys. B **64**, 189 (1986).

[2] N. F. Mott, *Metal Insulator transitions*, Taylor and Francis, London, 2nd edition, 1990.

[3] S. Jin, T. H. Tiefel, M. McCormack, R. A. Fastnacht, R. Ramesh, and L. H. Chen, *Thousandfold change in resistivity in magnetoresistive La-Ca-Mn-O films*, Science **264**, 413 (1994).

[4] D. D. Fong, G. B. Stephenson, S. K. Streiffer, J. A. Eastman, O. Auciello, P. H. Fuoss, and C. Thompson, *Ferroelectricity in ultrathin perovskite films*, Science **304**, 1650 (2004).

[5] J. Hubbard, *Electron correlations in narrow energy bands*, Proc. Phys. Soc., London, Sect. A **276**, 238 (1963).

[6] M. Imada, A. Fujimori, and Y. Tokura, *Metal-insulator transitions*, Rev. Mod. Phys. **70**, 1039 (1998).

[7] Y. Y. Wang, N. S. Rogado, R. J. Cava, and N. P. Ong, *Spin entropy as the likely source of enhanced thermopower in $Na_xCo_2O_4$*, Nature **423**, 425 (2003).

[8] H. A. Kramers, *The interaction between the magnetogenic atoms in a paramagnetic crystal*, Physica **1**, 182 (1934).

[9] C. Zener, *Interaction between the d-shells in the transition metals. II. Ferromagnetic compounds of manganese with perovskite structure*, Phys. Rev. **82**, 403 (1951).

[10] R. D. Shannon, *Revised effective ionic-radii and systematic studies of interatomic distances in halides and chalcogenides*, Acta Crystallogr. Sect. A **32**, 751 (1976).

[11] A. F. Wells, *Structural inorganic chemistry*, Oxford University Press, Oxford, 5th edition, 1991.

[12] A. M. Haghiri-Gosnet and J. P. Renard, *CMR manganites: physics, thin films and devices*, J. Phys. D **36**, R127 (2003).

[13] H. A. Jahn and E. Teller, *Stability of polyatomic molecules in degenerate electronic states. I. Orbital degeneracy*, Proc. Phys. Soc., London, Sect. A **161**, 220 (1937).

[14] J. H. Park, E. Vescovo, H. J. Kim, C. Kwon, R. Ramesh, and T. Venkatesan, *Direct evidence for a half-metallic ferromagnet*, Nature **392**, 794 (1998).

[15] J. A. Noland, *Optical absorption of single-crystal strontium titanate*, Phys. Rev. **94**, 724 (1954).

[16] A. Verneuil, *Memoiro on the artificial reproduction of ruby by fusion*, Ann. Chim. Phys. **3**, 20 (1904).

[17] H. Unoki and T. Sakudo, *Electron spin resonance of Fe^{3+} in $SrTiO_3$ with special reference to $110°K$ phase transition*, J. Phys. Soc. Jpn. **23**, 546 (1967).

[18] P. A. Fleury, J. F. Scott, and J. M. Worlock, *Soft phonon modes and $110°K$ phase transition in $SrTiO_3$*, Phys. Rev. Lett. **21**, 16 (1968).

[19] K. A. Müller, W. Berlinger, and F. Waldner, *Characteristic structural phase transition in perovskite-type compounds*, Phys. Rev. Lett. **21**, 814 (1968).

[20] N. Bickel, G. Schmidt, K. Heinz, and K. Müller, *Ferroelectric relaxation of the $SrTiO_3$ (100) surface*, Phys. Rev. Lett. **62**, 2009 (1989).

[21] V. Ravikumar, D. Wolf, and V. P. Dravid, *Ferroelectric-monolayer reconstruction of the $SrTiO_3$ (100) surface*, Phys. Rev. Lett. **74**, 960 (1995).

[22] J. F. Schooley, W. R. Hosler, and M. L. Cohen, *Superconductivity in semiconductiong $SrTiO_3$*, Phys. Rev. Lett. **12**, 474 (1964).

[23] M. Yoshimoto, T. Maeda, K. Shimozono, H. Koinuma, M. Shinohara, O. Ishiyama, and F. Ohtani, *Topmost surface-analysis of $SrTiO_3(001)$ by coaxial impact-collision ion-scattering spectroscopy*, Appl. Phys. Lett. **65**, 3197 (1994).

BIBLIOGRAPHY

[24] M. Kawai, Z. Y. Liu, T. Hanada, M. Katayama, M. Aono, and C. F. McConville, *Layer controlled growth of oxide superconductors*, Appl. Surf. Sci. **82** (1994).

[25] M. Kawasaki, K. Takahashi, T. Maeda, R. Tsuchiya, M. Shinohara, O. Ishiyama, T. Yonezawa, M. Yoshimoto, and H. Koinuma, *Atomic control of the $SrTiO_3$ crystal-surface*, Science **266**, 1540 (1994).

[26] G. Koster, B. L. Kropman, G. J. H. M. Rijnders, D. H. A. Blank, and H. Rogalla, *Quasi-ideal strontium titanate crystal surfaces through formation of strontium hydroxide*, Appl. Phys. Lett. **73**, 2920 (1998).

[27] T. Ohnishi, K. Shibuya, M. Lippmaa, D. Kobayashi, H. Kumigashira, M. Oshima, and H. Koinuma, *Preparation of thermally stable TiO_2-terminated $SrTiO_3$ substrate surfaces*, Appl. Phys. Lett. **85**, 272 (2004).

[28] R. Herger, P. R. Willmott, O. Bunk, C. M. Schlepütz, B. D. Patterson, and B. Delley, *Surface of strontium titanate*, Phys. Rev. Lett. **98**, 076102 (2007).

[29] R. Herger, P. R. Willmott, O. Bunk, C. M. Schlepütz, B. D. Patterson, B. Delley, V. L. Shneerson, P. F. Lyman, and D. K. Saldin, *Surface structure of $SrTiO_3(001)$*, Phys. Rev. B **76**, 195435 (2007).

[30] G. Charlton, S. Brennan, C. A. Muryn, R. McGrath, D. Norman, T. S. Turner, and G. Thornton, *Surface relaxation of $SrTiO_3(001)$*, Surf. Sci. **457**, L376 (2000).

[31] T. Matsumoto, H. Tanaka, T. Kawai, and S. Kawai, *STM-imaging of a $SrTiO_3(100)$ surface with atomic-scale resolution*, Surf. Sci. **278**, L153 (1992).

[32] Q. D. Jiang and J. Zegenhagen, *$c(6 \times 2)$ and $c(4 \times 2)$ reconstruction of $SrTiO_3(001)$*, Surf. Sci. **425**, 343 (1999).

[33] M. R. Castell, *Scanning tunneling microscopy of reconstructions on the $SrTiO_3(001)$ surface*, Surf. Sci. **505**, 1 (2002).

[34] V. Vonk, S. Konings, G. J. van Hummel, S. Harkema, and H. Graafsma, *The atomic surface structure of $SrTiO_3(001)$ in air studied with synchrotron x-rays*, Surf. Sci. **595**, 183 (2005).

[35] T. Kubo and H. Nozoye, *Surface structure of SrTiO$_3$(100)-($\sqrt{5} \times \sqrt{5}$)-R26.6°*, Phys. Rev. Lett. **86**, 1801 (2001).

[36] K. Szot and W. Speier, *Surfaces of reduced and oxidized SrTiO3 from atomic force microscopy*, Phys. Rev. B **60**, 5909 (1999).

[37] A. Kazimirov, D. M. Goodner, M. J. Bedzyk, J. Bai, and C. R. Hubbard, *X-ray diffraction analysis of structural transformations on the (001) surface of oxidzed SrTiO$_3$*, Surf. Sci. **492**, L711 (2001).

[38] N. Erdman, K. R. Poeppelmeier, O. Warschkow, D. E. Ellis, and L. D. Marks, *The structure and chemistry of the TiO$_2$-rich surface of SrTiO$_3$(001)*, Nature **419**, 55 (2002).

[39] F. Silly, D. T. Newell, and M. R. Castell, *SrTiO$_3$(001) reconstructions: the (2×2) to $c(4 \times 4)$ transition*, Surf. Sci. **600**, L219 (2006).

[40] J. E. T. Andersen and P. J. Møller, *Impurity-induced 900°C (2×2) surface reconstruction of SrTiO$_3$(100)*, Appl. Phys. Lett. **56**, 1847 (1990).

[41] Y. Liang and D. A. Bonnell, *Structures and chemistry of the annealed SrTiO$_3$(001) surface*, Surf. Sci. **310**, 128 (1994).

[42] E. Heifets, S. Piskunov, E. A. Kotomin, Y. F. Zhukovskii, and D. E. Ellis, *Electronic structure and thermodynamic stability of double-layered SrTiO$_3$(001) surfaces: Ab initio simulations*, Phys. Rev. B **75**, 115417 (2007).

[43] A. Ohtomo and H. Y. Hwang, *A high-mobility electron gas at the LaAlO$_3$/SrTiO$_3$ heterointerface*, Nature **427**, 423 (2004).

[44] G. H. Jonker and J. H. van Santen, *Ferromagnetic compounds of manganese with perovskite structure*, Physica **16**, 337 (1950).

[45] P. Schiffer, A. P. Ramirez, W. Bao, and S.-W. Cheong, *Low-temperature magnetoresistance and the magnetic phase-diagram of La$_{1-x}$Ca$_x$MnO$_3$*, Phys. Rev. Lett. **75**, 3336 (1995).

[46] A. Urushibara, Y. Moritomo, T. Arima, A. Asamitsu, G. Kido, and Y. Tokura, *Insulator-metal transition and giant magnetoresistance in La$_{1-x}$Sr$_x$MnO$_3$*, Phys. Rev. B **51**, 14103 (1995).

[47] C. Şen, G. Alvarez, H. Aliaga, and E. Dagotto, *Colossal magnetoresistance observed in Monte Carlo simulations of the one- and two-orbital models for manganites*, Phys. Rev. B **73**, 224441 (2006).

[48] A. J. Millis, P. B. Littlewood, and B. I. Shraiman, *Double exchange alone does not explain the resistivity of $La_{1-x}Sr_xMnO_3$*, Phys. Rev. Lett. **74**, 5144 (1995).

[49] E. Dagotto, T. Hotta, and A. Moreo, *Colossal magnetoresistant materials: The key role of phase separation*, Phys. Rep. **344**, 1 (2001).

[50] W. A. Harrison, *Electronic structure and the properties of solids: the physics of the chemical bond*, Dover Publications, Inc., New York, 1989.

[51] J. B. Goodenough, *Theory of the role of covalence in the perovskite-type manganites $[La,M(II)]MnO_3$*, Phys. Rev. **100**, 564 (1955).

[52] P. W. Anderson and H. Hasegawa, *Considerations on double exchange*, Phys. Rev. **100**, 675 (1955).

[53] F. Rivadulla, M. Otero-Leal, A. Espinosa, A. de Andres, C. Ramos, J. Rivas, and J. B. Goodenough, *Suppression of ferromagnetic double exchange by vibronic phase segregation*, Phys. Rev. Lett. **96**, 016402 (2006).

[54] P.-G. de Gennes, *Effects of double exchange in magnetic crystals*, Phys. Rev. **118**, 141 (1960).

[55] G.-M. Zhao, K. Conder, H. Keller, and K. A. Müller, *Giant oxygen isotope shift in the magnetoresistive perovskite $La_{1-x}Ca_xMnO_{3+y}$*, Nature **381**, 676 (1996).

[56] S. Mori, C. H. Chen, and S. W. Cheong, *Pairing of charge-ordered stripes in $(La,Ca)MnO_3$*, Nature **392**, 473 (1998).

[57] M. Fäth, S. Freisem, A. A. Menovsky, Y. Tomioka, J. Aarts, and J. A. Mydosh, *Spatially inhomogeneous metal-insulator transition in doped manganites*, Science **285**, 1540 (1999).

[58] M. Uehara, S. Mori, C. H. Chen, and S.-W. Cheong, *Percolative phase separation underlies colossal magnetoresistance in mixed-valent manganites*, Nature **399**, 560 (1999).

[59] A. Moreo, S. Yunoki, and E. Dagotto, *Phase separation scenario for manganese oxides and related materials*, Science **283**, 2034 (1999).

[60] T. Hotta, A. Feiguin, and E. Dagotto, *Stripes induced by orbital ordering in layered manganites*, Phys. Rev. Lett. **86**, 4922 (2001).

[61] Y. Tokura and Y. Tomioka, *Colossal magnetoresistive manganites*, J. Magn. Magn. Mater. **200**, 1 (1999).

[62] E. O. Wollan and W. C. Koehler, *Neutron diffraction study of the magnetic properties of the series of perovskite-type compounds $[(1-x)La, xCa]MnO_3$*, Phys. Rev. **100**, 545 (1955).

[63] C. Şen, G. Alvarez, and E. Dagotto, *Competing ferromagnetic and charge-ordered states in models for manganites: The origin of the colossal magnetoresistance effect*, Phys. Rev. Lett. **98**, 127202 (2007).

[64] Y. Lu, X. W. Li, G. Q. Gong, G. Xiao, A. Gupta, P. Lecoeur, J. Z. Sun, Y. Y. Wang, and V. P. Dravid, *Large magnetotunneling effect at low magnetic fields in micrometer-scale epitaxial $La_{0.67}Sr_{0.33}MnO_3$ tunnel junctions*, Phys. Rev. B **54**, R8357 (1996).

[65] S. Yuasa, T. Nagahama, A. Fukushima, Y. Suzuki, and K. Ando, *Giant room-temperature magnetoresistance in single-crystal Fe/MgO/Fe magnetic tunnel junctions*, Nature Mat. **3**, 868 (2004).

[66] C. H. Ahn, K. M. Rabe, and J.-M. Triscone, *Ferroelectricity at the nanoscale: Local polarization in oxide thin films and heterostructures*, Science **303**, 488 (2004).

[67] C. H. Ahn, A. Bhattacharya, M. Di Ventra, J. N. Eckstein, C. D. Frisbie, M. E. Gershenson, A. M. Goldman, I. H. Inoue, J. Mannhart, A. J. Millis, A. F. Morpurgo, D. Natelson, and J.-M. Triscone, *Electrostatic modification of novel materials*, Rev. Mod. Phys. **78**, 1185 (2006).

[68] K. Dörr, *Ferromagnetic manganites: spin-polarized conduction versus competing interactions*, J. Phys. D **39**, R125 (2006).

Chapter 3

Theory

3.1 Diffraction

In this section, a short introduction to the basic principles of x-ray diffraction of crystals in general and surfaces in particular is given. The reader is referred to textbooks or reviews for a more detailed description [1–5].

3.1.1 Single crystal diffraction

The interaction between x-rays and crystals can assumed to be weak and hence, no change in intensity and energy due to (multiple) scattering or absorption is seen. This is called the *kinematical approximation* and offers a simplified theoretical description of the interaction between x-rays and crystalline material compared to other sources of radiation, e.g., electrons, for which the assumption is no longer valid and one has to treat the interactions *dynamically*.

In surface x-ray diffraction (SXRD), the scattering process is usually treated elastically, i.e., within the kinematical approximation. However, inelastic scattering on a periodic structure, e.g., Compton scattering, contributes to the background. The high-intensity regions close to Bragg reflections would require a dynamical treatment, but since they mainly contain information about the bulk and not the surface, they are usually excluded from consideration.

In what follows, some of the most important aspects and formulae of diffraction within the kinematical limit are given.

The crystal lattice of an ideal, infinitely extended, periodic crystal has a unit cell, spanned by the three lattice vectors **a**, **b** and **c**. So any lattice point site in real space can be described by the vector

$$\mathbf{R}_n = n_1\mathbf{a} + n_2\mathbf{b} + n_3\mathbf{c}, \quad (3.1)$$

where n_1, n_2 and n_3 are integers. Within each unit cell the atoms are on positions $\mathbf{r}_j$. The volume V of the unit cell is given by $V = \mathbf{a} \cdot (\mathbf{b} \times \mathbf{c})$. The corresponding reciprocal space[1] is spanned by the three reciprocal base vectors (indicated by the superscripted $*$):

$$\mathbf{a}^* = 2\pi\frac{\mathbf{b} \times \mathbf{c}}{V}, \quad \mathbf{b}^* = 2\pi\frac{\mathbf{c} \times \mathbf{a}}{V}, \quad \mathbf{c}^* = 2\pi\frac{\mathbf{a} \times \mathbf{b}}{V}. \quad (3.2)$$

Every point in reciprocal space can be described as a linear combination of the three reciprocal lattice vectors. One talks of a reciprocal lattice site when the linear combination consists only of integer multiples of the reciprocal lattice vectors.

An incoming, monochromatic, electromagnetic beam of wavelength λ has a wavevector $\mathbf{k}$, with $|\mathbf{k}| = 2\pi/\lambda$, and the outgoing scattered wave has a wavevector $\mathbf{k}'$, $\mathbf{k}$ and $\mathbf{k}'$ enclosing an angle of 2θ. Although we only deal with elastic scattering of x-rays, momentum can be transferred. We therefore introduce the scattering vector $\mathbf{q}$, defined as

$$\hbar\mathbf{q} = \hbar\mathbf{k}' - \hbar\mathbf{k}, \quad (3.3)$$

where $\hbar\mathbf{k}$ and $\hbar\mathbf{k}'$ are the initial and final momenta of the photon, respectively.

Starting from classical Thomson theory of scattering from a single electron, one can derive, that for diffraction on an ideal single crystal, intensity can only be measured on discrete points in reciprocal space. These are the so-called Bragg points. Three equivalent formulations are used to describe this: the Laue equations

$$\mathbf{q} \cdot \mathbf{a} = 2\pi h, \quad \mathbf{q} \cdot \mathbf{b} = 2\pi k, \quad \mathbf{q} \cdot \mathbf{c} = 2\pi l, \quad (3.4)$$

the Ewald construction

$$\mathbf{q} = h\mathbf{a}^* + h\mathbf{b}^* + h\mathbf{c}^*, \quad (3.5)$$

and the Bragg law

$$2d_{hkl}\sin\theta = n\lambda. \quad (3.6)$$

The integers h, k and l are the Miller indices. The distance between (hkl) planes is given by d_{hkl}, and, e.g., for a crystal with a cubic unit cell of length a this equates to $d_{hkl} = a/\sqrt{h^2 + k^2 + l^2}$.

3.1. DIFFRACTION

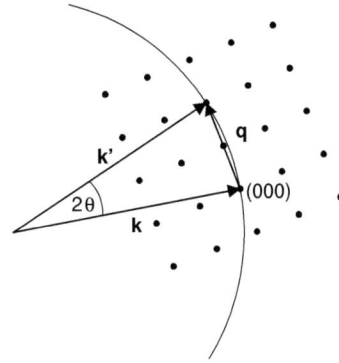

Figure 3.1: The Ewald construction. The scattering vector $\mathbf{q} = \mathbf{k'} - \mathbf{k}$ must begin at the (0 0 0) diffraction spot of the incident beam, and end at another diffraction maximum, the reciprocal lattice point. As x-ray diffraction is an elastic process, this means that the two points must be on a sphere of radius $|\mathbf{k}| = 2\pi/\lambda$ and thus defining the value of θ.

n is a positive integer. The Ewald construction is shown in Fig. 3.1 and is one possibility to illustrate x-ray diffraction as an elastic process. An important consequence is that the scattering vector $\mathbf{q}$ always lies perpendicular to the scattering planes of the crystal.

The intensity at a Bragg point I_{hkl} is the square of the modulus of the *complex structure factor* F_{hkl}: $I_{hkl} \propto |F_{hkl}|^2$. For a scattering vector $\mathbf{q}$, the structure factor is given by

$$F(\mathbf{q}) = \sum_j f_j(|\mathbf{q}|)\exp(i\mathbf{q} \cdot \mathbf{r}_j), \tag{3.7}$$

where f_j are the *atomic form factors* of the electrons, i.e., the Fourier transformations of the electron distribution of an atom. It approaches the atomic number of the atom Z for $\mathbf{q} \to 0$ and becomes 0 for $\mathbf{q} \to \infty$. The sum in Eqn. 3.7 includes all atoms of a unit cell. The concept of the structure factor will be resumed after the introduction of the convolution theorem in the next section.

We now square Eqn 3.7 to obtain

$$|F_{hkl}|^2 = F_{hkl}F_{hkl}^* = F_{hkl}F_{\overline{hkl}} = (F_{\overline{hkl}})^*F_{\overline{hkl}} = |F_{\overline{hkl}}|^2. \tag{3.8}$$

---
[1] also called *k*-space

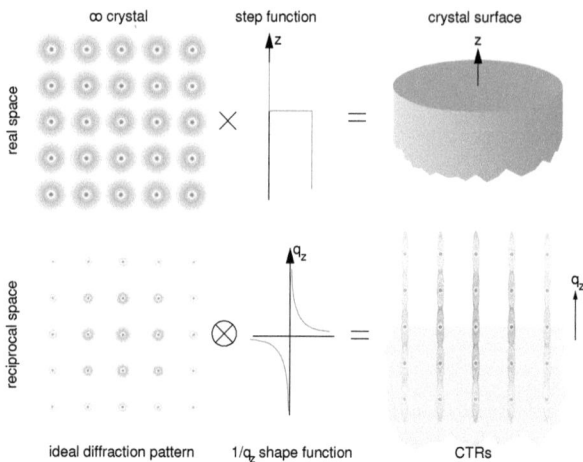

Figure 3.2: The generation of crystal truncation rods explained by the help of the convolution theorem. Only half of the Friedel pairs can be probed (see Eqn. 3.8), because the surface shadows scattering below the horizon (shown here as CTR signal on the shaded background).

This is Friedel's rule, connecting F_{hkl} and $F_{\overline{hkl}}$ to so-called Friedel pairs. This is only true within the kinematical limit, i.e., far from absorption edges.

3.1.2 Crystal truncation rods

The diffraction pattern of an ideal, infinite, 3-dimensional (3D) crystal produces a set of δ-functions, the Bragg peaks. Only when the scattering vectors **q** connects two reciprocal lattice points [i.e., the (0 0 0) and the desired one], one can measure intensity. The truncation of this crystal with a surface makes it a semi-infinite, 2-dimensional (2D) system. The condition is relaxed and the peaks are smeared out in the dimension perpendicular to the truncated surface. The intensity variation along this direction is termed a *crystal truncation rod* (CTR).

For understanding CTRs, it is important to introduce the *convolution theorem*. It states that the Fourier transform (FT) of a convolution of two functions f and g is equal to the product of their individual FTs. Mathematically (for simplified notation only in 1D, although also valid

3.1. DIFFRACTION

in 3D), this can be described as

$$\text{FT}[f(x) \otimes g(x)] = \text{FT}[f(x)] \, \text{FT}[g(x)], \qquad (3.9)$$

with the convolution ($\otimes$) defined as

$$f(x) \otimes g(x) = \int_{-\infty}^{+\infty} f(x')g(x-x')dx'. \qquad (3.10)$$

Hence, the diffraction pattern of an ideal, truncated crystal can be generated by convoluting the FT of an infinitely large crystal structure (i.e., the set of δ-functions) with the FT of the function describing the boundary of the crystal, the shape function. Figure 3.2 illustrates the application of the convolution theorem to surfaces. Here, the related shape function in the surface normal is a step function. Its FT has a $1/z$-relationship that extends significantly in k-space. Away from a Bragg reflection, the scattering intensity is thus proportional to $1/q_z$ and the intensity to $1/q_z^2$. Hence again, the effect of the surface is to smear out the intensity along the direction perpendicular to that surface.

It is important to note that in most cases the boundary function is irregular on an atomic scale and consequently its FT is very narrow. Therefore, effects other than this (such as crystallographic imperfections, beam divergence and monochromaticity) will determine the linewidths of the diffraction peaks.

We now turn to the structure factor again. One can think of the electronic distribution of an ideal crystal as the convolution of the Bravais lattice (i.e., a 3D set of δ-functions, also referred to as "comb function") with the electronic distribution of the basis, i.e., within the unit cell. The FT of a 3D set of δ-functions is another 3D set of δ-functions with separations inversely proportional to the separations of the original array. Using the convolution theorem, the diffraction pattern of an infinite crystal is the product of the FT of the comb function of the Bravais lattice and the FT of the electronic distribution within a unit cell. This latter FT is the *structure factor* (see also Eqn. 3.7) and therefore determines the intensities of the peaks in the diffraction pattern.

So far, we did not take into account any deviations of the structure factor due to thermal movements of the atoms (even at $T = 0$ K, due to a finite zero point energy as a consequence of Heisenberg's uncertainty principle) or spatial inhomogeneities of the structure. For a given temperature one cannot distinguish between these two different sorts of deviations of the averaged positions. One introduces therefore the so-called Debye-Waller factor of the form

$$M = \frac{1}{2}\mathbf{q}^T \frac{\mathbf{B}_j}{8\pi^2}\mathbf{q}, \qquad (3.11)$$

with the dispersion matrix $\mathbf{B}_j/(8\pi^2)$, which is a symmetric (3×3) matrix. The superscript T indicates the transposed matrix of $\mathbf{q}$, i.e., here the corresponding row vector. Sometimes, one calls also $\exp(-M)$ the DW factor, but we keep with the definition given in Eqn. 3.11 throughout this work. For isotropic DW factors, we can define the root mean-square-displacement of an atom j as

$$\sigma_j = \sqrt{\frac{B_j}{8\pi^2}}. \qquad (3.12)$$

The complete structure factor of a unit cell accounting for atomic displacements (thermal and spatial) plus reflecting the possibility of an occupancy of less than unity at a site j, θ_j, is then given by

$$F(\mathbf{q}) = \sum_j \theta_j \exp\left(-\frac{1}{2}\mathbf{q}^T \frac{\mathbf{B}_j}{8\pi^2}\mathbf{q}\right) f_j(|\mathbf{q}|) \exp(i\mathbf{q}\cdot\mathbf{r}_j). \qquad (3.13)$$

Note that we did not ask for the presence of a surface in Eqn. 3.13.

Finally, we derive an expression of the intensity variation along a CTR, i.e., we consider the scattering along the surface normal $\mathbf{c}$. The scattering amplitude of a semi-infinite stack of n layers composed of unit cells with structure factors $F(\mathbf{q})$ (as defined in Eqn. 3.13) can be written as

$$F_{\text{ctr}}(\mathbf{q}) = F(\mathbf{q}) \sum_{n=-\infty}^{0} \exp(iq_z nc) = \frac{F(\mathbf{q})}{1-\exp(i2\pi l)}, \qquad (3.14)$$

where the wavevector transfer perpendicular to the surface is $q_z = 2\pi l/c$. The intensity along a CTR is obtained by squaring Eqn. 3.14:

$$I_{\text{ctr}}(\mathbf{q}) \propto |F_{\text{ctr}}(\mathbf{q})|^2 = \frac{|F(\mathbf{q})|^2}{4\sin^2(\pi l)}. \qquad (3.15)$$

Thus, the intensity variation along l proportional to $1/\sin^2(\pi l)$ for unreconstructed and relaxed surfaces.

3.2 Superstructure rods

A commensurable surface reconstruction has a fixed phase relation to the underlying bulk unit cell. The unit cell vectors of the reconstruction, $\mathbf{a}_{\text{rcn}}$ and $\mathbf{b}_{\text{rcn}}$, can therefore simply be expressed as a linear combination of the surface cell vectors $\mathbf{a}_{\text{surf}}$ and $\mathbf{b}_{\text{surf}}$:

$$\begin{pmatrix} \mathbf{a}_{\text{rcn}} \\ \mathbf{b}_{\text{rcn}} \end{pmatrix} = \mathbf{G} \begin{pmatrix} \mathbf{a}_{\text{surf}} \\ \mathbf{b}_{\text{surf}} \end{pmatrix} = \begin{pmatrix} g_{11} & g_{12} \\ g_{21} & g_{22} \end{pmatrix} \begin{pmatrix} \mathbf{a}_{\text{surf}} \\ \mathbf{b}_{\text{surf}} \end{pmatrix}. \qquad (3.16)$$

3.3. DIRECT METHODS

In the general case of a commensurable reconstruction, it is labeled by stating the matrix **G**. For a case with $g_{12} = g_{21} = 0$, the notation is simplified to $(g_{11} \times g_{22})$. This is used, for example, in the case of the (2×1) and (2×2) reconstructions of strontium titanate (STO, see Papers I and II). If the reconstruction is rotated by angle α, but conserving the angle between the cell vectors **a** and **b**, then the notation is $(|\mathbf{a}_{\text{rcn}}| \times |\mathbf{b}_{\text{rcn}}|)R\alpha$. An example could be the STO(001)-$\binom{2\ 1}{1\ 2}$ reconstruction, that is then denoted as STO(001)-$(\sqrt{5} \times \sqrt{5})R26.6°$.

Reconstructed surfaces, where the surface unit cell is bigger than the bulk unit cell, do give rise to superstructure rods (SSRs). *SSRs do not contain any bulk information.* The structure factor of an SSR can be derived analogously to the bulk structure factor (see Eqn. 3.13). Again, the measured intensity is proportional to the structure factor:

$$I_{\text{ssr}}(\mathbf{q}) \propto |F_{\text{ssr}}(\mathbf{q})|^2. \tag{3.17}$$

The FT of a δ-function is a constant. If all atoms of the reconstruction are at the same z-position, the structure factor without considering the DW factor is then constant, i.e., independent of l. Taking the DW factor into account, the intensity decreases monotonically along l. Variations of the atomic positions in z or reconstructed surfaces do characteristically modulate this intensity profile, making qualitative judgments of the reconstruction sometimes possible. The total number of scattering atoms of a surface reconstruction is much smaller than that for a bulk crystal, therefore leading to much weaker diffraction signals that can often only be reliably recorded at synchrotron radiation sources.

A commensurate $(m \times n)$ surface reconstruction has reciprocal unit cell size of $(1/m \times 1/n)$. The scattered signal coming from the reconstruction can therefore produce SSRs alone, or SSRs superimposed on CTRs, depending on the position in k-space. The total signal along these rods is composed of the coherent superposition of SSR and CTR signals, as follows:

$$I_{\text{tot}}(\mathbf{q}) \propto |F_{\text{tot}}(\mathbf{q})|^2 = \left| F_{\text{ctr}}(\mathbf{q}) + \frac{\theta_{\text{ssr}}}{A_{\text{ssr}}} F_{\text{ssr}}(\mathbf{q}) \right|^2, \tag{3.18}$$

where θ_{ssr} is the fraction of the surface that is reconstructed and A_{ssr} is the area of the reconstructed unit cell, thus normalized to the (1×1) bulk unit cell.

3.3 Direct methods

One of the fundamental problems of crystallography becomes evident when either Eqn. 3.7 or Eqn. 3.13 is squared in order to get an intensity: One looses the phase information. Several

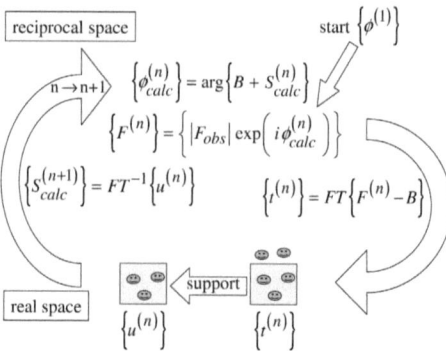

Figure 3.3: The iterative phase recovery algorithm PARADIGM alternatively satisfies constraints in real and reciprocal space and exploits the fact that scattering from the unknown surface structure may be regarded as a perturbation of that from the truncated bulk structure. Courtesy of P.F. Lyman.

attempts have been made to overcome this obstacle, mostly for bulk diffraction. In this section, two possible solutions to solve this problem for SXRD are discussed. The first is the so-called PARADIGM approach by D.K. Saldin's group in Milwaukee, Wisconsin, USA, with which we established a fruitful collaboration that had significant impact on solving the structure of STO. In the upcoming Section 3.3.2, a second solution called COBRA (in an equally successful collaboration with R. Clarke, Ann Arbor, Michigan, USA and Y. Yacoby, Jerusalem, Israel) will be presented, as it was used for the structure determination of ultrathin films of lanthanum strontium manganate.

3.3.1 PARADIGM

The PARADIGM algorithm aims to recover the surface structure by an iterative algorithm initially proposed by Fienup [6]. The algorithm is detailed elsewhere [7], however, a brief description is given below. In Fig. 3.3, a schematic flowchart of the iterative algorithm is shown (reproduced from Ref. [8]).

Initially, a random set of phases[2] ϕ is assigned to the experimentally observed structure

---
[2] The curly brackets in Fig. 3.3 underline that complete sets and not individual elements of a set are meant. However, in the text the notation is simplified and the brackets omitted, although the meaning remains the same.

3.3. DIRECT METHODS

factor amplitudes $|F_{\text{obs}}|$. The bulk contribution B is calculated and subtracted. A FT generates a real space estimate of the electron density t in the surface region. Next, a physical reasonable constraint is applied: the electron density must lie within a spatial region of a few Ångströms across the surface, outside, the electron density is set to zero. The constrained electron density of the unknown surface contribution, u, is inverse-Fourier-transformed into k-space, yielding a set of complex surface structure factors S_{calc} that can be added to the calculated bulk contribution B. The arguments of the sums represent improved estimates of the phases ϕ. A constraint in reciprocal space assigns these phases to the experimentally observed structure factor amplitudes $|F_{\text{obs}}|$, and the cycle is repeated until the error between S_{calc} and F is minimized and a predefined convergence criterion is met.

3.3.2 COBRA

COBRA stands for Coherent Bragg Rod Analysis and is another recently developed phase-retrieval algorithm [9, 10]. COBRA uses the fact that the complex structure factor (CSF) along the CTR has to be continuous, so the phase has to be, as well. The COBRA method is generally applicable to systems that are periodic in two dimensions, aperiodic in the third, and commensurate with the underlying substrate.

The electron density (ED) is represented as the sum of a reference structure ED and an unknown difference ED, describing the deviations of the surface structure to the bulk structure. The same approach holds true for k-space, where the total CSF is the sum of the reference and unknown CSFs. Considering two adjacent points along a rod separated by $\Delta \mathbf{k}$, we can write the total CSF as

$$S\left(\mathbf{k}-\frac{\Delta \mathbf{k}}{2}\right)+U\left(\mathbf{k}-\frac{\Delta \mathbf{k}}{2}\right) = T\left(\mathbf{k}-\frac{\Delta \mathbf{k}}{2}\right),$$
$$S\left(\mathbf{k}+\frac{\Delta \mathbf{k}}{2}\right)+U\left(\mathbf{k}+\frac{\Delta \mathbf{k}}{2}\right) = T\left(\mathbf{k}+\frac{\Delta \mathbf{k}}{2}\right), \quad (3.19)$$

where S, U, and T are the CSFs due to the reference, unknown, and total electron densities, respectively.

Now, we make use of the fact that the CSFs vary continuously along CTRs and make the approximation that at two adjacent points along a rod,

$$U\left(\mathbf{k}-\frac{\Delta \mathbf{k}}{2}\right) \cong U\left(\mathbf{k}+\frac{\Delta \mathbf{k}}{2}\right) = U_a(\mathbf{k}). \quad (3.20)$$

This approximation will be accurate if U indeed does vary slower than S. This is ensured by a combination of two means: First, the reference ED is similar to that of the real system, so their

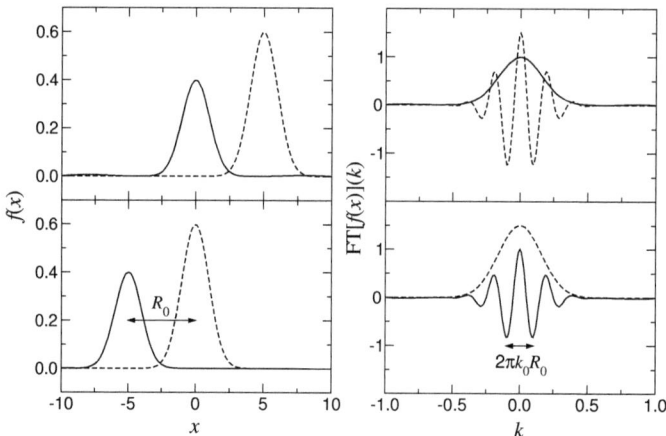

Figure 3.4: Two Gaussian line shapes (left) and their FTs (right) with different origins. The solid-line Gaussian represents the unknown part U in Eqn 3.19, while the reference part S of the CSF is given by the dashed-line. Displacing the origin in real space by a vector $\mathbf{R}_0$ changes the phase of the FT of the electron density by a factor $\exp(i\mathbf{k}\cdot\mathbf{R}_0)$. Thus, choosing the origin of the real space coordinate close to the unknown part causes the phase rate of change of U to vary more slowly than that of S.

CSFs are of the same order of magnitude. Second, the real space origin is chosen to be close to the top of the thin film so that the CSF of the reference structure will vary more rapidly along the rods than that of the unknown structure (see Fig. 3.4).

If we now take the modulus of Eqn. 3.19 and introduce the COBRA approximation (Eqn. 3.20), we obtain for the CSF (i.e., a quantity proportional to the experimentally observed intensity):

$$\left| S\left(\mathbf{k}-\frac{\Delta\mathbf{k}}{2}\right)+U_a(\mathbf{k})\right| = \left|T\left(\mathbf{k}-\frac{\Delta\mathbf{k}}{2}\right)\right|,$$
$$\left| S\left(\mathbf{k}+\frac{\Delta\mathbf{k}}{2}\right)+U_a(\mathbf{k})\right| = \left|T\left(\mathbf{k}+\frac{\Delta\mathbf{k}}{2}\right)\right|. \quad (3.21)$$

This yields two real equations that can be solved for one complex unknown. In general, this pair of equations has two solutions (U_a, U_b) and it is necessary to choose the correct one.

In Fig. 3.5, a graphical solution of Eqn. 3.21 is presented for two consecutive pairs of adjacent points, $\mathbf{k}-\Delta\mathbf{k}/2$ and $\mathbf{k}+\Delta\mathbf{k}/2$ [(a), corresponding complex numbers marked with

3.3. DIRECT METHODS

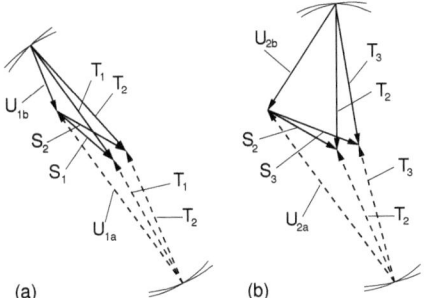

Figure 3.5: Graphical solution of Eqn. 3.21 of two consecutive pairs of adjacent points. (a) The solutions U_{1a} and U_{1b} for the points at $\mathbf{k} - \Delta\mathbf{k}/2$ and $\mathbf{k} + \Delta\mathbf{k}/2$. (b) The solutions U_{2a} and U_{2b} at $\mathbf{k} + \Delta\mathbf{k}/2$ and $\mathbf{k} + 3\Delta\mathbf{k}/2$. The difference between U_{1a} and U_{2a} is smaller than the difference between any other pair of solutions, so this is the correct pair.

indices 1 and 2, respectively] or $\mathbf{k} + \Delta\mathbf{k}/2$ and $\mathbf{k} + 3\Delta\mathbf{k}/2$ [(b), corresponding complex numbers marked with indices 2 and 3, respectively]. Each pair of equation has two solutions U_a and U_b shown as dashed and solid lines, respectively. Under the assumption that U varies slowly along the CTRs, the correct pair of solutions are those which change the least when going from one point to the next, i.e., U_{1a} and U_{2b} in Fig. 3.5. This procedure reveals the unknown CSFs and hence the total CSFs. To obtain the ED, the CSFs are Fourier-transformed into real space.

The iterative COBRA cycles are shown in Fig. 3.6. In the resulting ED, unphysical residues, such as negative ED are removed. The application of this constraint also affects the structure factors. Therefore, the restrained ED is Fourier-transformed for comparison with the measured intensities. If the deviations are still too large, a further iteration must be performed with the restrained ED as the new reference structure. This process is iterated until convergence is reached.

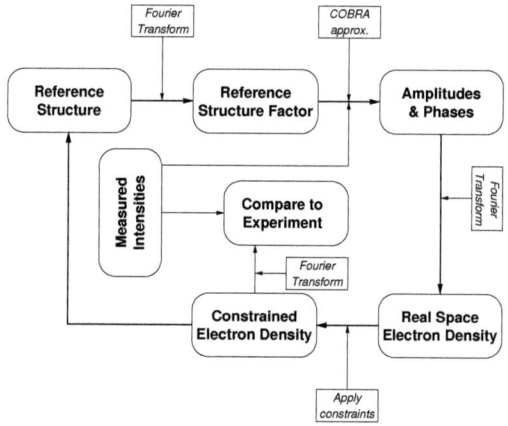

Figure 3.6: Flow diagram of the COBRA data analysis procedure.

Bibliography

[1] B. E. Warren, *X-ray diffraction*, Dover Publications, Inc., New York, 1990.

[2] C. Giacovazzo, H. L. Monaco, G. Artioli, D. Viterbo, G. Ferraris, G. Gilli, G. Zanotti, and M. Catti, *Fundamentals of crystallography*, Oxford University Press Inc., New York, 2nd edition, 2002.

[3] J. Als-Nielsen and D. McMorrow, *Elements of modern x-ray physics*, John Wiley and Sons Ltd., New York, 2001.

[4] R. Feidenhans'l, *Surface structure determination by x-ray diffraction*, Surf. Sci. Rep. **10**, 105 (1989).

[5] I. K. Robinson and D. J. Tweet, *Surface x-ray diffraction*, Rep. Prog. Phys. **55**, 599 (1992).

[6] J. R. Fienup, *Phase retrieval algorithms - a comparison*, Appl. Opt. **21**, 2758 (1982).

[7] D. K. Saldin, R. J. Harder, V. L. Shneerson, and W. Moritz, *Phase retrieval methods for surface x-ray diffraction*, J. Phys. Condens. Matter **13**, 10689 (2001).

[8] P. F. Lyman, V. L. Shneerson, R. Fung, R. J. Harder, E. D. Lu, S. S. Parihar, and D. K. Saldin, *Atomic-scale visualization of surfaces with x-rays*, Phys. Rev. B **71**, 081402(R) (2005).

[9] M. Sowwan, Y. Yacoby, J. Pitney, R. MacHarrie, M. Hong, J. Cross, D. A. Walko, R. Clarke, R. Pindak, and E. A. Stern, *Direct atomic structure determination of epitaxially grown films: Gd_2O_3 on GaAs(100)*, Phys. Rev. B **66**, 205311 (2002).

[10] Y. Yacoby, M. Sowwan, E. Stern, J. Cross, D. Brewe, R. Pindak, J. Pitney, E. B. Dufresne, and R. Clarke, *Direct determination of epitaxial film and interface structure: Gd_2O_3 on GaAs (100)*, Physica B **336**, 39 (2003).

Chapter 4

Experiment

4.1 Pulsed laser deposition

4.1.1 Setup

Two different ultra-high vacuum-compatible (UHV) setups were used for thin film production. The *ex-situ* chamber was primarily used for optimization of the growth conditions, and was equipped with a 20 keV electron gun to perform reflection high-energy electron-diffraction (RHEED).

The *in-situ* chamber can be mounted directly on the diffractometer of the SD station at the MS BL and provides a small entrance and a large exit Be-windows. The chamber has been detailed elsewhere [1]. Schematics of the *ex-situ* and the *in-situ* setups are shown in Fig. 4.1(a) and (b), respectively. Both chambers contain a pulsed valve that allows one to supplement the growth by synchronizing a reactive gas jet to the laser pulse. For LSMO thin film growth, N_2O was chosen as the primary oxidant due to its enhanced oxidizing properties compared to O_2 [2, 3].

A 12.7 mm diameter ablation rod composed of highly pure, polycrystalline sections of $LaMnO_3$ (LMO) and $SrMnO_3$ (SMO), manufactured by Praxair Surface Technologies, Woodinville WA, was used as the laser target. Contamination due to adsorbents such as water or CO_2 were removed by preablating the target before deposition.

The laser radiation was the fourth harmonic (4ω) of a pulsed Nd:YAG[1] laser (Quantel Brilliant B, 266 nm wavelength, 5 ns pulse length, 10 Hz repetition rate) entering through a Brew-

---
[1] YAG = yttrium aluminium garnet

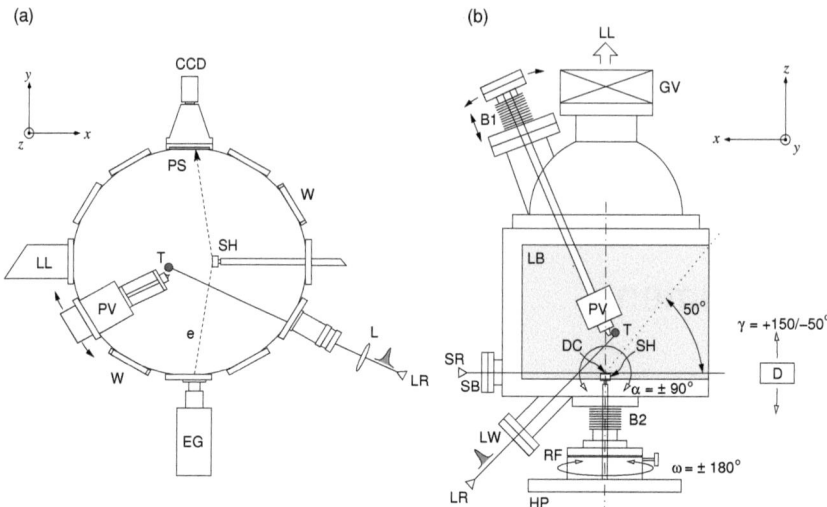

Figure 4.1: Plan projection from above on the PLD chambers. The *ex-situ* chamber is shown in (a). LL = loadlock; PS = phosphorescence screen; CCD = charge coupled device camera; W = window; SH = substrate manipulator arm with sample holder and substrate; L = lens with $f = 650$ nm; LR = laser radiation; EG = electron gun; e = electron beam; T = target; PV = pulsed valve. (b) shows the *in-situ* chamber. LL = loadlock; GV = gate valve to loadlock; B1, B2 = edge-welded bellows; LB = large Be-window; PV = pulsed valve; DC = diffractometer center; SH = substrate heater/holder; SR = synchrotron radiation; SB = small Be-window; D = detector; LW = Brewster laser entrance window; RF = rotation feedthrough for substrate rotation; HP = hexapod front plate; LR = laser radiation. Diffractometer movements are given by open arrows.

ster window into the chamber. The energy per pulse is controlled by a variable attenuator, reducing it to about 24 mJ/pulse. The focused laser spot had a size of about 1 mm² and hence the fluence was about 2 J cm^{-2}.

4.1.2 Substrate preparation

Strontium titanate (SrTiO$_3$, STO) (001) substrates (space group $Pm\bar{3}m$, $a_0 = 3.9045$ Å) with low vicinality ($< 0.1°$) and an impurity content of < 5 ppm Ca [4] were purchased from CrysTec GmbH, Germany. Substrates had dimensions of $10 \times 8 \times 1$ mm³ (*in-situ*) or $10 \times$

4.1. PULSED LASER DEPOSITION

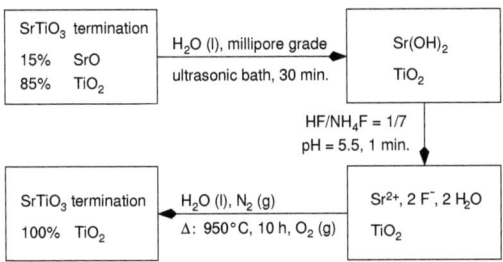

Figure 4.2: Preparation process in order to obtain TiO_2-terminated STO, according to Refs. [5, 6].

10×0.5 mm^3 (*ex-situ*). They were prepared according to an established chemical and thermal treatment [5, 6], in order to ensure 100% termination on the TiO_2 atomic layer and smooth terrace edge profiles.

Briefly, the preparation involves the following steps (see Fig. 4.2). As-received STO was immersed in very pure water (millipore grade, electrical resistivity of $\geq$ 18.2 MΩ cm) and ultrasounded for 30 min in order to activate the surface. TiO_2 remains inert whereas the SrO forms hydroxide complexes at the surface. Etching the STO in buffered HF (BHF, HF/NH$_4$F = 1/7, pH $\approx$ 5.5, Riedel-de-Haën) for 1 min removed the hydroxide complexes and a TiO_2-terminated surface was preserved. Residuals of the BHF were washed away by extensive rinse in millipore water. The clean surfaces were dried in a high quality N$_2$ jet (grade 5.0) over several minutes and subsequently annealed in a tube furnace in 1 atm of pure O$_2$ (grade 5.0) for 10 hours at a temperature of 950 °C. The furnace was then turned off and cooled down.

4.1.3 Growth

The ablation of La$_{1-x}$Sr$_x$MnO$_3$ (LSMO) took place in an O$_2$ background of 2×10^{-2} Pa and the synchronized N$_2$O gas pulse of 1.5×10^{-2} Pa average pressure. Note that due to the pulsed nature of the N$_2$O jet, the instantaneous gas pressure interacting with the ablation species is much higher, enhancing the oxidation process even more [7].

The substrates were ohmically heated to 700 °C using a 0.375 mm thick Si wafer, sandwiched in between the holder and substrate by clamping. Temperature control was achieved by comparing the resistivity [typically $R(700\,°C) = 1.2\,\Omega$, i.e., $U = 5.3$ V, $I = 4.4$ A] to a calibration curve. Occasional cross-checks were made using the pyrometer mounted at the *ex-situ* chamber.

In order to grow the alloy $La_{1-x}Sr_xMnO_3$, the ratio x of LMO/SMO could be adjusted by the choice of the ablation boundaries using a high axial translation velocity of rod (typically $v_t \approx 20$ mm s^{-1}) and taking the different ablation ratio of LMO and SMO into account, as described elsewhere [8].

4.2 Surface x-ray diffraction

Surface x-ray diffraction (SXRD) experiments were performed at the at the Surface Diffraction station (SD) station at the Materials Science Beamline at the Swiss Light Source, Paul Scherrer Institut, Villigen, Switzerland. The beamline has been detailed elsewhere [9].

4.2.1 Diffractometer

The SD station makes use of a $(2+3)$-circle diffractometer [10]. A schematic showing the diffractometer, the *in-situ* PLD and two detectors types (i.e., point and area detector) is shown in Fig. 4.3. The diffractometer is controlled by the shell-based software SPEC.

It is noted that the diffractometer underwent several redesign steps since the setup shown in Fig. 4.3, although this schematic expresses most accurately the setup that was used for taking the main part of the data presented here. The most prominent improvements are: (i) the new Pilatus 100k pixel detector (replacing the old Pilatus I, used for collecting most of the data in this work), and (ii), that the pixel detector is mounted directly on the v-axis, replacing the point detector.

4.2.2 Detector

For the reasons given above, the Pilatus I detector system is described here. The commercially available Pilatus 100k detector that is nowadays in use at the MS BL is detailed elsewhere [11].

The Pilatus I detector contains an array of $366 \times 157 = 57462$ pixels, each with dimensions of 217×217 μm^2, covering an active area of 7.9×3.4 cm^2. The distance from the center of diffraction to the detector was 1.235 m. Assuming an infinitely small sample size, each pixel covered an angle of $0.0101°$, hence the whole the detector subtended a solid angle of $3.70°$

4.2. SURFACE X-RAY DIFFRACTION

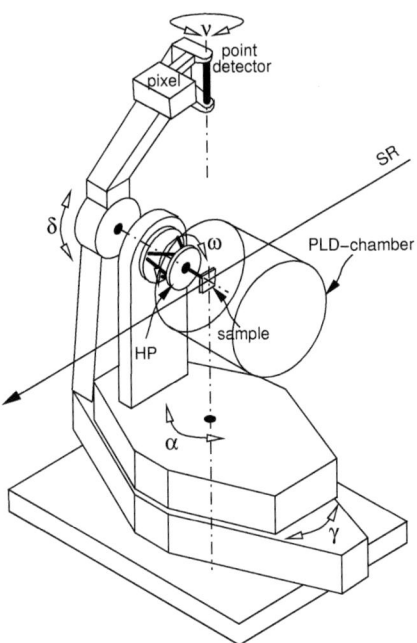

Figure 4.3: The $(2+3)$-circle surface diffractometer. There are two degrees of freedom for sample movements, α and ω, and three for moving the detector, γ, δ, and ν. The PLD chamber sits on the α-table. Fine adjustment of the in-vacuum sample surface relative to the diffractometer center and incoming synchrotron radiation (SR) is achieved by the hexapod (HP).

$\times 1.59°$. This is enough area to capture an entire SXRD signal and its surrounding background, and facilitates the identification of artifacts.

Each pixel has a 15-bit counter, i.e., a high dynamic range (2^{15}) [2] compared to that for a charge-coupled device (CCD) of typically $\sim 2^8$, and the whole detector can be read out in < 7 ms [12]. High frame rates, however, could never been achieved because the network was not optimized for high data throughput. This problem is resolved with the new detector.

Further advantages of this new detector technology are the excellent point spread function (i.e., only the pixel hit by a photon really counts a signal) and the zero readout noise. In combi-

---
[2] but still one and a half orders of magnitude less than the Pilatus 100k with 2^{20}!

Figure 4.4: A typical Pilatus I image after the application of standard correction factors according to Ref. [14]: The data was flat-field corrected in order to account for different counting efficiencies of the individual pixels, and dead pixels were replaced by the median value of the counts of the eight nearest-neighbor pixels. The image comprises of the diffraction signal of the (2 1) CTR at $l = 1.05$ (feature to the left) and the foot of the (2 1 1) Bragg peak (right) of $SrTiO_3$.

nation with the significantly improved dynamic range (compared to CCDs), this makes Pilatus pixel detectors ideally suited for recording signal strongly varying in intensity (e.g., SXRD data) and/or extremely dense packed diffraction patterns, as observed in protein crystallography.

4.2.3 Reciprocal space scans

The traditional way of recording SXRD data is to record a rocking scan for each point along a CTR using a point detector. In doing so, the intensity for each l value is collected by rotating the sample about its surface normal (i.e., around ω), subtracting a background and integrating the signal. Thanks to the availability of the Pilatus area detectors, the intensity can be recorded in a faster way.

In the so-called "stationary mode" [13], one records the complete diffraction signal in one single exposure, and, thus, no sample rotation is needed for capturing the signal of a specific l point in reciprocal space. The use of this mode accelerates data taking by a factor of at least an order of magnitude, and, in combination with the larger field of view of the area detector in k-space, also the reliability is improved. A typical Pilatus I image with standard correction factors applied is shown in Fig. 4.4 [14].

4.2. SURFACE X-RAY DIFFRACTION

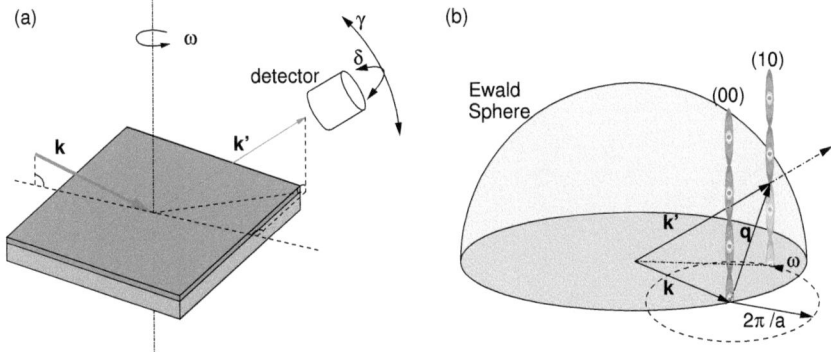

Figure 4.5: Measurement principle of crystal truncation rods. (a) The experimental geometry in real space including the paths of incident **k** and scattered beam **k'**, as well as the three relevant diffractometer angles γ, δ and ω are shown. (b) Rotation of the sample around its normal axis ω moves the (1 0) rod around the specular [(0 0)] rod in reciprocal space and through the Ewald sphere.

In order to record intensities along the l-direction of a CTR using the stationary mode, one moves the detector to the wanted position in k-space and records an image. A visualization of the measurement principle in real and reciprocal space is given in Fig. 4.5.

The stationary mode, however, has to be used with care. Although the detector acceptance is no longer a problem, the outgoing angle β_{out} has to be large enough to integrate at once. This limit depends on the angle of incidence, the out-of-plane lattice constant, the quality of the crystal and the desired resolution Δl with with the rod shall be sampled (see Fig. 4.6) [13]. For any given resolution, the minimum outgoing angle is defined as

$$\tan \beta_{out} \geq \frac{2\Delta Q_{FWHM}}{c^* \Delta l}, \qquad (4.1)$$

where ΔQ_{FWHM} is the full-width-half-maximum of a CTR and c^* the reciprocal out-of-plane lattice unit. With respect to the substrate lattice, this value corresponds to a minimum l_{min} value, above which independent structure factor amplitudes are achieved. In the case of STO, typical in-plane rocking curves had widths of $\Delta Q_\parallel = 0.0044$ Å^{-1} (i.e., $\Delta \omega = 0.02°$), a value which was subsequently used in combination with the wanted sampling resolution Δl to determine l_{min}.

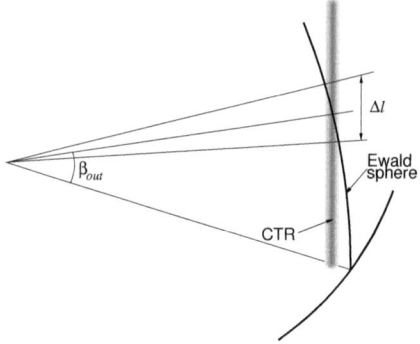

Figure 4.6: A CTR cutting through the Ewald sphere illustrating the dependence of the outgoing angle β_{out} and the sampling resolution Δl.

4.3 Atomic force microscopy

Atomic force microscopy (AFM) in tapping mode was performed using a Veeco Dimension 3100. The AFM tip in use was a MikroMasch NSC15/Al BS with a characteristic frequency of about 300 kHz. Samples were measured in air. The scanned area was typically 1 μm^2.

4.4 X-ray photoemission spectroscopy

X-ray photoelectron spectroscopy (XPS) was performed under UHV conditions using an Al Kα source at 1487 eV and a Gammadata Scienta SES 2002 hemispherical electron analyzer. The escape depth of the photoelectrons is some 25 Å for Al Kα radiation [15].

We measured the angle-dependent electron yield of the oxygen 1s (O1s) signal of STO in order to characterize the surface termination. STO is an insulator (band gap = 3.2 eV) [16] and causes charging problems. Thus, only the relative shifts in energy, i.e., the chemical shifts, could be reliably established. The change of the coordination number of Sr (Ti) from 12 (6) in the bulk to 8 (5) at the surface has a direct influence on the chemical shift of the O1s signal and can therefore be used as a fingerprint to identify the surface termination.

4.5. RUTHERFORD BACKSCATTERING

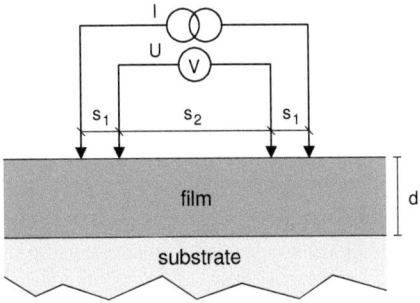

Figure 4.7: A schematic of the four-point method in linear electrode geometry for measuring the resistivity of thin films. s_1, s_2 = electrode distances; d = film thickness.

4.5 Rutherford backscattering

The thin films were analyzed for their stoichiometry using Rutherford backscattering. The measurements were performed using a 2 MeV ^{4}He beam and a silicon surface barrier detector at a scattering angle of 165° [17]. The background has been subtracted using a recently developed fitting procedure [18]. The data were analyzed using the RUMP program [19]. For very thin films with a thickness of the order of several MLs, however, the RUMP analysis fails, but the stoichiometry could be determined by element-specific integration of the backscattering signal and correlation to thicker films grown under identical conditions.

4.6 Resistivity

The resistance R of the LSMO samples was measured in a Quantum Design Model 6000 physical properties measurement system (PPMS) using two different contact methods to measure the voltage drop. The films were connected either via gold wires and silver paste or a recently developed sample holder design, where four springed pins contact the surface without the need of metal paste. Both designs gave reproducible results.

The resistivity ρ depends on the measurement geometry, but can be directly related to the ohmic resistance $R = U/I$ using appropriate correction factors. Figure 4.7 shows a typical electrode setup. In order to obtain ρ, one has to integrate the effective electric field across the electrodes that gives rise to the voltage drop along s_2. For bulk samples, radial symmetry of the field $E(r) \propto 1/r^2$ is usually assumed. For thin films, however, the film thickness is constant $(r \gg d$, i.e., $E(r) \propto 1/r)$ and the resistivity can be described by

$$\rho = \pi R d \ln^{-1}\left(\frac{s_1 + s_2}{s_1}\right). \tag{4.2}$$

The sample holder thus has a correction factor of $g = (s_1 + s_2)/s_1 = 6$ ($g = \sqrt{2}$) for the linear (quadratic) setup, whereas the silver pasted electrodes are approximated with $g \approx 2$.

The magnetic field in the PPMS was applied normal to the film surface and varied from 0 to 7 T.

Bibliography

[1] P. R. Willmott, C. M. Schlepütz, B. D. Patterson, R. Herger, M. Lange, D. Meister, D. Maden, C. Brönnimann, E. F. Eikenberry, G. Hülsen, and A. Al-Adwan, *In situ studies of complex PLD-grown films using hard x-ray surface diffraction*, Appl. Surf. Sci. **247**, 188 (2005).

[2] P. Lecoeur, A. Gupta, P. R. Duncombe, G. Q. Gong, and G. Xiao, *Emission studies of the gas-phase oxidation of Mn during pulsed laser deposition of manganates in O_2 and N_2O atmospheres*, J. Appl. Phys. **80**, 513 (1996).

[3] P. R. Willmott, R. Herger, M. C. Falub, L. Patthey, M. Döbeli, C. V. Falub, M. Shi, and M. Schneider, *Pulsed laser deposition of atomically flat $La_{1-x}Sr_xMnO_3$ thin films using a novel target geometry*, Appl. Phys. A **79**, 1199 (2004).

[4] R. Herger and K. Peters, private communication with CrysTec GmbH, 2007.

[5] M. Kawasaki, K. Takahashi, T. Maeda, R. Tsuchiya, M. Shinohara, O. Ishiyama, T. Yonezawa, M. Yoshimoto, and H. Koinuma, *Atomic control of the $SrTiO_3$ crystal-surface*, Science **266**, 1540 (1994).

[6] G. Koster, B. L. Kropman, G. J. H. M. Rijnders, D. H. A. Blank, and H. Rogalla, *Quasi-ideal strontium titanate crystal surfaces through formation of strontium hydroxide*, Appl. Phys. Lett. **73**, 2920 (1998).

[7] P. R. Willmott and J. R. Huber, *Pulsed laser vaporization and deposition*, Rev. Mod. Phys. **72**, 315 (2000).

[8] P. R. Willmott, R. Herger, and C. M. Schlepütz, *Multilayers, alloys, and complex profiles by pulsed laser deposition using a novel target geometry*, Thin Solid Films **453-54**, 436 (2004).

[9] B. D. Patterson, R. Abela, H. Auderset, Q. Chen, F. Fauth, F. Gozzo, G. Ingold, H. Kühne, M. Lange, D. Maden, D. Meister, P. Pattison, T. Schmidt, B. Schmitt, C. Schulze-Briese, M. Shi, M. Stampanoni, and P. R. Willmott, *The Materials Science Beamline at the Swiss Light Source: design and realization*, Nucl. Instrum. Methods A **540**, 42 (2005).

[10] E. Vlieg, *A $(2+3)$-type surface diffractometer – mergence of the z-axis and $(2+2)$ geometries*, J. Appl. Crystallogr. **31**, 198 (1998).

[11] Dectris, Next generation x-ray detectors, http://www.dectris.ch.

[12] C. Brönnimann, S. Florin, M. Lindner, B. Schmitt, and C. Schulze-Briese, *Synchrotron beam test with a photon-counting pixel detector*, J. Synchr. Rad. **7**, 301 (2000).

[13] E. Vlieg, *Integrated intensities using a six-circle surface x-ray diffractometer*, J. Appl. Crystallogr. **30**, 532 (1997).

[14] C. M. Schlepütz, R. Herger, P. R. Willmott, B. D. Patterson, O. Bunk, C. Brönnimann, B. Henrich, G. Hülsen, and E. F. Eikenberry, *Improved data acquisition in grazing-incidence x-ray scattering experiments using a pixel detector*, Acta Crystallogr. Sect. A **61**, 418 (2005).

[15] M. P. Seah and W. A. Dench, *Quantitative electron spectroscopy of surfaces: A standard data base for electron inelastic mean free paths in solids*, Surf. Interface Anal. **1**, 2 (1979).

[16] J. A. Noland, *Optical absorption of single-crystal strontium titanate*, Phys. Rev. **94**, 724 (1954).

[17] M. Döbeli, R. M. Ender, V. Liechtenstein, and D. Vetterli, *Time-of-flight spectrometry applied to 2 MeV He RBS*, Nucl. Instrum. Methods B **142**, 417 (1998).

[18] M. Döbeli, *Subtraction tool for the analysis of RBS data*, Nucl. Instrum. Methods B **249**, 800 (2006).

[19] L. R. Doolittle, *A semiautomatic algorithm for Rutherford backscattering analysis*, Nucl. Instrum. Methods B **15**, 227 (1986).

Chapter 5

Concluding remarks

Transition metal oxides with the perovskite structure are an interesting class of material, due to the complex interplay of the different electronic and structural degrees of freedom, leading to fascinating physical effects. The explanation of these are in most cases not trivial: Only when the full picture arising from experiments and theory is considered, can one expect to be able to give a qualitative *and* quantitative description.

Because of the strong influences of subtle structural changes on the electronic properties of such strongly correlated electron systems, it can be expected that heterogeneous surfaces and interfaces will show different physical properties than those of the bulk. We have found this to be the case in the work presented here. For example, the investigations of the surface of strontium titanate showed that the sensitivity of the surface to the preparation and processing conditions can cause unexpected effects, such as a terminating TiO_2-double layer; the semiconducting response of the resistivity of lanthanum strontium manganate revealed the presence of a dead layer although the magnetoresistive properties were conserved.

The work on the ultrathin manganite films showed that atomic control of crystal growth down to a fraction of a unit cell can be achieved using state-of-the-art equipment. The subsequent *in-situ* recording of surface x-ray diffraction using the novel Pilatus area pixel detector demonstrates one of the many unique possibilities that research using synchrotron radiation can offer. However, such an experiment would not have been possible without thorough *ex-situ* characterization beforehand and afterwards. It is hoped that the combination of different techniques to study one (class of) material(s) will be further developed and standardized, and thus materials scientists become motivated to study materials of increasing complexity.

The kinetic studies demonstrated the promise of pulsed laser deposition as the premier tool to achieve controlled growth on the atomic level. Its relative experimental simplicity makes pulsed laser deposition not only ideally suited as a versatile research tool but also opens the path towards industrial applications for nanoscaled devices.

Real space techniques, such as scanning tunneling microscopy, offer direct access to a crystal surface which in turn can simplify the interpretation. However, probing local areas in real space always raises the question of representativeness. k-space techniques on the other hand, usually reveal the system's averaged property. If one is interested in this, they can complement the real space information and reduce the problem of the spatially limited field of view, but the subsequent interpretation is often not straightforward. In surface crystallography, the loss of the phase information excludes the direct Fourier transformation of the diffraction pattern into electron densities. The recent development of direct methods applicable to crystal surfaces helps to overcome this drawback. In particular, the increasing complexity of the studied systems makes conventional structure refinement based on model-guessing by trial-and-error and minimization of a goodness of fit criterion very inefficient and time-consuming. A realistic starting model for structure refinement can be extracted by first applying phase-retrieval methods on the experimental diffraction data. This not only minimizes the time to solve a structure, but also improves the reliability of the result. By having good analytic tools at hand to estimate the phases, surface x-ray diffraction offers access to atomic length scales with unprecedented accuracy in combination with a unique depth resolution normal to the surface of a crystal.

The results of a quantitative surface x-ray diffraction analysis are very accurate atomic positions. These can be exceedingly helpful for the explanation of the physical properties of the material. The work on lanthanum strontium manganate gives an indication of the possible potential of this: the refined atomic positions could be used to determine a lower limit for the onset of magnetoresistance. The detailed knowledge of the structure of the thin film monolayer-by-monolayer can help to model the system as consisting of three basic regions interface, film and surface.

Knowing exactly the atomic positions is on the one hand essential for theoretical investigations such as density functional theory calculations. On the other hand, it is an important step towards tunable and nanoscaled devices in technological applications. The control of magnetoresistance by mechanical stress or strain, or by chemical doping via substitution could improve the versatility, e.g., to applications in different temperature regimes. The combination of functional oxides with different physical properties in intimately connected heterostructures could lead to a breathtaking downsizing of the technical component to length scales of only a few unit cells!

An increasingly unified understanding of transition metal oxides exhibiting exciting physical properties; the breakthrough in synthesis and materials processing techniques; standardization of characterization tools capable of treating highly complex oxides; and the advances made in the theoretical description of these systems lead the author to believe that transition metal oxide compounds will have a similarly prosperous future as that enjoyed by semiconductors, and that they will impact on all sorts of technological applications in our civilization.

Paper I
Surface of strontium titanate

The work presented in this chapter has been published in:
R. Herger, P.R. Willmott, O. Bunk, C.M. Schlepütz, B.D. Patterson, and B. Delley, *Surface of strontium titanate*, Phys. Rev. Lett. **98**, 076102 (2007).

artikel1.pdf is the online version of the library (Zentralbibliothek Zürich).

Abstract

We report the first complete determination, using surface x-ray diffraction, of the surface structure of TiO_2-terminated $SrTiO_3(001)$, both at room temperature in vacuum, and also hot, under typical conditions used for thin film growth. The room-temperature structure consists of a mixture of a (1×1) relaxation and (2×1) and (2×2) reconstructions. The latter disappear over several minutes upon heating. The structures are best modeled by a TiO_2-rich surface similar to that proposed by Erdman *et al.* [Nature (London) **419**, 55, (2002).]. Both reconstructions have been shown by density functional theory to be energetically favorable. The calculated surface energy of the (1×1) relaxation is higher, indicating that it may be a disordered mixture of the reconstructions. Displacements of the atoms from their bulk positions are significant down to three unit cells, which may have important implications on possible surface ferroelectric phenomena in $SrTiO_3$.

DOI: 10.1103/PhysRevLett.98.076102 [a]
PACS numbers: 68.35.Bs, 61.10.Nz, 71.15.Mb, 77.84.Bw

Reprinted with kind permission from the American Physical Society.
[a] Note that you need a subscription for this journal to directly access the article.

Surface of Strontium Titanate

R. Herger, P. R. Willmott,* O. Bunk, C. M. Schlepütz, and B. D. Patterson
Swiss Light Source, Paul Scherrer Institut, CH-5232 Villigen, Switzerland

B. Delley
Condensed Matter Theory Group, Paul Scherrer Institut, CH-5232 Villigen, Switzerland
(Received 22 November 2006; published 15 February 2007)

We report the first complete determination, using surface x-ray diffraction, of the surface structure of TiO_2-terminated $SrTiO_3(001)$, both at room temperature in vacuum, and also hot, under typical conditions used for thin film growth. The cold structure consists of a mixture of a (1×1) relaxation and (2×1) and (2×2) reconstructions. The latter disappear over several minutes upon heating. The structures are best modeled by a TiO_2-rich surface similar to that proposed by Erdman *et al.* [Nature (London) **419**, 55 (2002).]. Both reconstructions have been shown by density functional theory to be energetically favorable. The calculated (1×1) surface energy is higher, indicating that it may be a disordered mixture of the reconstructions. Atomic displacements are significant down to three unit cells, which may have important implications on possible surface ferroelectric phenomena in $SrTiO_3$.

DOI: 10.1103/PhysRevLett.98.076102   PACS numbers: 68.35.Bs, 61.10.Nz, 71.15.Mb, 77.84.Bw

Strontium titanate ($SrTiO_3$, STO) is the paradigmatic substrate material for epitaxial growth of thin films of the technologically important family of perovskites, including high-temperature superconductors, colossal magnetoresistive materials, ferroelectrics, and heterostructures containing two-dimensional electron gases [1–4]. The surface structure of STO has therefore been the subject of intense research in recent years and a detailed description of the surface with sub-Å resolution is hence of great importance. Despite this effort, the surface structure of STO remains a subject of controversy, due on the one hand to its sensitivity to the preparation and processing conditions, and on the other, to the limited spatial resolution and chemical sensitivity associated with most surface characterization techniques. According to preparation and ambient conditions, at least 8 different reconstructions and relaxations have been reported for the STO(001) surface alone [5–14].

Among these, the unusual (2×1) reconstruction proposed by Erdman *et al.* [5] has provoked much debate [9,11,12,14,15]. In this model, the surface terminates not with one, but two TiO_2 layers, i.e., the surface is Ti rich, while the topmost Ti atoms cluster to form a "zigzag" motif. Such structural distortions and rearrangements could affect the nucleation mechanisms in thin film growth and may also induce different physical properties from those of the bulk, including ferroelectricity [2,16], catalytic behavior, and spin-polarization/double exchange mechanisms at thin film interfaces to manganites [17].

For thin film growth, a reliable chemical and thermal preparation of the STO substrate surface has become established, which ensures 100% termination on the TiO_2 sublayer (SL) and smooth terrace edge profiles [18–20]. In this Letter we present the temperature-dependent structure of STO(001) prepared in this standard manner, first at room temperature ("cold") and in ultrahigh vacuum (UHV), and, subsequently, under conditions typical for perovskite thin film growth (heated to $750 \pm 30\,°C$ in 10^{-3} Pa O_2—"hot"). The structures were determined using surface x-ray diffraction (SXRD) [21,22], which is one of only a few techniques capable of providing the sub-Å resolution necessary to predict the detailed physical properties of crystalline surfaces. The different surface structures presented here were found to all exhibit Ti clustering and a surface TiO_2 double layer.

The morphology of the prepared STO substrate surface was checked using atomic force microscopy (AFM) and found to consist only of straightedged terraces of 0.4 nm height (i.e., one unit cell) and 250 nm width, while x-ray photoelectron spectroscopy confirmed termination with TiO_2. Substrates were then introduced in a UHV chamber equipped with a large beryllium window [23] mounted on a $(2 + 3)$ circle surface diffractometer at the materials science beam line of the Swiss Light Source, Paul Scherrer Institut. Thanks to the availability of a fast x-ray photon-counting pixel detector [24], it was feasible to reliably record the large data set of independent structure factors necessary to obtain a convincing model within the limited allocated beam time. Two independent sets of data from two different samples were recorded with 1 Å synchrotron radiation for both the cold and hot conditions, to confirm reproducibility.

In addition to crystal truncation rods (CTRs), superstructure rods (SSRs) associated with (2×2) and (2×1) surface reconstructions were identified in the SXRD signal for the cold sample. These SSR and CTR signals were stable in the cold conditions over the measurement time of over 24 h, though the SSRs vanished over several minutes when the substrate was brought to the hot conditions [25].

A representative subset of the cold data is shown in Fig. 1(b), along with the error bars, which derive primarily from systematic differences between symmetry-equivalent structure factors. The complete data set consists of 9 CTRs and 18 SSRs, totaling 1668 nonequivalent structure factors, (plus 806 equivalent data points) and spans $|h|$, $|k|$, and $l = 0$ to 3. The fits were generated by the program FIT [26] using 394 parameters (i.e., an oversampling factor of 4). These included atomic coordinates, Debye-Waller (DW) factors, and fractional contributions from each surface-structure type. Although the final goodness of fit is given in terms of the crystallographic R factor, fit optimization was via χ^2-minimization, in order to avoid artificial weighting of the most intense signal near the Bragg maxima in the CTRs [27].

The structure comprises all 3 surface domains shown in Fig. 1(a), including their symmetry-equivalent orientations, and significant relaxations are observed down to a depth of 3 unit cells. The final model, shown in Fig. 1(c), has a crystallographic R factor of 4.5% [28] and no unphysical positions or DW factors. It is dominated by the (2×2) (43%) and (2×1) (37%) reconstructions, plus a smaller contribution (20%) from a (1×1) relaxation.

The model shown in Fig. 1 was only one of 50 that were tested for the cold data, including selected models from the literature [5,6,9,11,13,29,30]. In arriving at this model, we were guided by clear and consistent trends: all models which contained both the surface TiO$_2$ double layer (DL) and a zigzag motif of the top Ti-atoms in the (2×1) and (2×2) [5,14,30] produced significantly better fits than other models. A detailed description of all investigated models will be given elsewhere [31].

The SXRD data of STO(001), recorded under the hot conditions, consisted of 837 structure factors plus 764 equivalent data points ($|h|$, $|k|$, and $l = 0$ to 3) and showed no SSRs. The data was fit with 57 atomic positional and DW parameters, i.e., an oversampling factor of more than 10. Seven models were tested. The best final surface structure contained no unphysical parameters and is shown in Fig. 2. It has a crystallographic R factor of 11.2%. It is very similar to the DL (1×1) structure for the cold data, but with more puckering of the top TiO$_2$ SL, as the topmost Ti atom moves further out from the surface.

The physical correctness of the models is supported on the one hand by previous experimental evidence regarding the DL (2×1) domain [5], but also by theoretical calculations, which predicted the low surface energy of STO(001) terminated in much the same manner as the DL (2×2) described here [30]. Indeed, our own density functional theory (DFT) calculations [32,33] using the PBE functional [34] have shown that the three lowest surface energy configurations of STO that were tested include the two reconstructions we have found experimentally (see Table I). The DL (2×2) surface can be considered as consisting of alternately flipped DL (2×1) surface cells, and hence their similar chemistry explains their comparable surface energies.

The other low-energy configuration is the bulk (1×1) surface (i.e., that without the TiO$_2$ overlayer). Fits of the data using this bulk (1×1) termination produced significantly poorer results. Also, this surface would be half a unit cell lower or higher (± 0.2 nm) than adjacent DL reconstructions, and yet we see no evidence of this in the AFM images. It has been suggested in the literature that such "ideal" bulk (1×1) surfaces do not exist in practice, due to surface oxygen vacancies [12]. Although we have no evidence of such vacancies from our fits, the slightly enlarged Debye-Waller factors of the surface oxygen atoms may reflect less than unity occupation.

Transformation between the DL (1×1) relaxation and either reconstruction only requires a diagonal hop of every second Ti-atom across half a surface (1×1) unit cell. However, the surface energy of the TiO$_2$ DL (1×1) seems

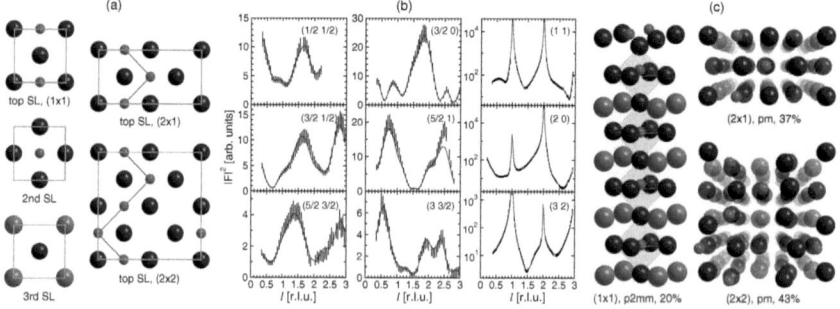

FIG. 1 (color online). The surface structure of cold STO: (a) Starting models of the 3 different domains. SL = sublayer. The 2nd SL (TiO$_2$) and 3rd SL (SrO) together make up a bulk STO unit cell. Note the zigzag motifs in the (2×1) and (2×2) reconstructions highlighted in red. (b) Subset of the SXRD data (black) and fits (red). We have included the $(5/2\ 2\ 1l)$ SSR, which shows the largest deviations between the fit and experimental data for the entire set. (c) The final models for the three domains, including their symmetries and percentage contributions. The (1×1) structure is viewed from the side, while the reconstructions are from above. Blue, O; red, Ti; green, Sr.

TABLE I. Surface energies in eV/(1 × 1) unit cell (with respect to the SrO chemical potential) of 4 domain types calculated using DFT. The bulk (1 × 1) model does not contain the extra TiO_2 sublayer shown upper left in Fig. 1(a).

Bulk (1 × 1)	DL (1 × 1)	DL (2 × 1)	DL (2 × 2)
0.78	1.39	0.78	0.38

anomalously high (Table I). The presence, in spite of this, of a (1 × 1) structure could have two reasons: either this domain is metastable with a significant activation barrier to lower-energy states, or in fact consists of a (possibly dynamic) disordered matrix of the zigzag motif of the (2 × 1) and (2 × 2), randomly flipped and mirrored [35]. This disordered (1 × 1) domain would have a surface energy comparably low to those of the two reconstructions, and the average structure is adequately described by the (1 × 1) structure shown in Fig. 1. Such a model would not have to invoke coordinated and concerted Ti hopping across the surface over large distances (of the order of 100 nm) to go from the reconstructions to the (1 × 1) relaxation. Disorder could also explain how such a (1 × 1) surface can dominate at elevated temperatures, where the surface thermal vibrational energy becomes comparable to the activation barrier between the (2 × 1) and (2 × 2), leading to a disordered mixture of these. Indeed, the difference in surface energy between two (2 × 1) cells and a single (2 × 2) cell is $\Delta E_s = 1.6$ eV (see Table I). On the other hand, the vibrational energy of the surface atoms of the same system is, to a first approximation, equal to $E_v = 3kT(4n_a)$, where k is Boltzmann's constant, T is the temperature, n_a is the number of atoms per top sublayer and unit cell (here, for TiO_2, $n_a = 3$), and the factor 4 accounts for the fact that 4 surface (1 × 1) cells are needed to describe this system. We therefore obtain $E_v = 3.2$ eV under hot conditions, i.e., twice the difference in reconstructions' surface energies. The size of the activation barrier between the two reconstructions will affect the time needed to reach this mixed equilibrium state. From the Arrhenius rate constant of the disappearance of the SSR signal of the order of $k_s = 0.01$ s^{-1} and assuming a preexponential factor $A \sim 10^{12}$, we obtain an activation energy of $E_a \approx 3$ eV. For the room temperature sample, $E_v = 0.9$ eV and the reaction rate constant is of the order of 10^{-40} s^{-1}; i.e., the system is completely kinetically hindered [36].

A further argument supporting the DL model is based on the electrostatic requirements of polar surfaces. The partially covalent nature of the bonds in STO results in the sublayers of the bulk (TiO_2 and SrO) having nonzero and opposite net charges $\pm\sigma$ [37]. For such polar systems, the top sublayer of stable surfaces are required to have a charge of $\sigma/2$, in order to avoid a physically unreasonable linear increase in the electric field with depth into the crystal. This condition is met by the top TiO_2 sublayer in bulk-terminated STO(001), which differs from lower TiO_2 sublayers by having a reduced number of bonds. The addition of the extra TiO_2 sublayer in the DL model modifies the charge of the second TiO_2 sublayer. However, simple calculations show that this TiO_2 overlayer in all three proposed DL surface structures compensates for this change and satisfies the criterion of electrostatic compensation.

Displacements, Δz, of the atomic positions in the final models compared to the high-symmetry positions in the starting models are shown in Fig. 3 in the direction of the sample normal, as it is in this direction that they are most prominent. These data, derived from experimental results, are very similar in trend to those theoretically predicted by Johnston et al. [12], with the uppermost atoms on average

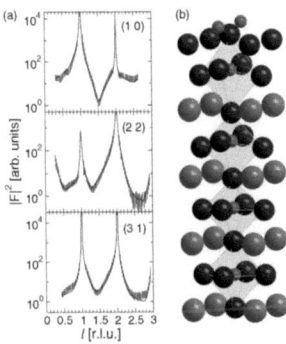

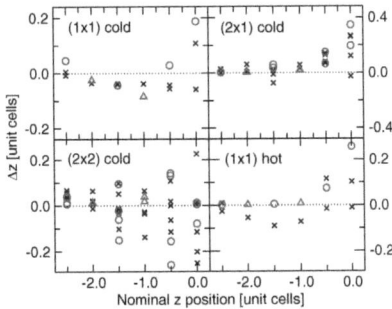

FIG. 2 (color online). The surface structure of (1 × 1) hot STO: (a) Subset of three rods from a total of nine of the SXRD data (black) and their fits (red), based on the same starting model as the DL (1 × 1) domain for the cold data shown in Fig. 1(a). (b) The final model for the hot DL (1 × 1) surface ($p2mm$ symmetry). Color code as in Fig. 1.

FIG. 3 (color online). Displacements of Sr (green triangles), Ti (red circles), and O (blue crosses) in the z direction (i.e., normal to the surface) from the high-symmetry positions in the starting models, as a function of z for the proposed cold and hot models. The nominal surface is at $z = 0$ and positive values of Δz indicate displacements towards the vacuum.

displaced outwards to the vacuum. The loss of centrosymmetry of the Ti atoms within the oxygen octahedra may lead to a permanent surface dipole moment and surface ferroelectricity [2,16,38]. Indeed, surface polarity is tacitly implied in the electrostatic arguments given above.

In conclusion, we have solved the surface structure of the scientifically and technologically important perovskite material $SrTiO_3(001)$ using surface x-ray diffraction, under optimal preparation and typical thin film growth conditions. For the first time, 3 structures simultaneously present on the same surface could be refined from a single data set. All three structures contain a characteristic double TiO_2 top layer, while the two reconstructions, predicted by DFT calculations to be energetically favorable, are formed by repetition of a common zigzag motif. It is suggested that the (1×1) structure may be an energetically favorable, disordered mixture of the two reconstructions. This would help explain the presence of only the (1×1) structure upon heating the sample, as the surface vibrational energy exceeds the difference in surface energy between the (2×1) and (2×2) domains, resulting in their complete mixing. The final models exhibit significant deviations from their high-symmetry starting positions down to a depth of 3 unit cells, which may have important consequences on the surface's ferroelectric and other nonlinear properties. Finally, it is hoped that theoretical calculations based on these models will provide a deeper insight into the physics of this fascinating surface.

We thank D. K. Saldin, P. F. Lyman, and V. L. Shneerson for fruitful discussions. Support of this work by the Schweizerischer Nationalfonds zur Förderung der wissenschaftlichen Forschung and the staff of the Swiss Light Source is gratefully acknowledged. This work was performed at the Swiss Light Source, Paul Scherrer Institut.

*Electronic address: philip.willmott@psi.ch
[1] S. Jin, T.H. Tiefel, M. McCormack, R.A. Fastnacht, R. Ramesh, and L. H. Chen, Science **264**, 413 (1994).
[2] D. D. Fong, G. B. Stephenson, S. K. Streiffer, J. A. Eastman, O. Auciello, P. H. Fuoss, and C. Thompson, Science **304**, 1650 (2004).
[3] C. H. Ahn, K. M. Rabe, and J.-M. Triscone, Science **303**, 488 (2004).
[4] A. Ohtomo and H. Y. Hwang, Nature (London) **427**, 423 (2004).
[5] N. Erdman, K. R. Poeppelmeier, O. Warschkow, D. E. Ellis, and L. D. Marks, Nature (London) **419**, 55 (2002).
[6] T. Matsumoto, H. Tanaka, T. Kawai, and S. Kawai, Surf. Sci. **278**, L153 (1992).
[7] Q. D. Jiang and J. Zegenhagen, Surf. Sci. **425**, 343 (1999).
[8] G. Charlton, S. Brennan, C. A. Muryn, R. McGrath, D. Norman, T. S. Turner, and G. Thornton, Surf. Sci. **457**, L376 (2000).
[9] T. Kubo and H. Nozoye, Phys. Rev. Lett. **86**, 1801 (2001).
[10] A. Kazimirov, D. M. Goodner, M. J. Bedzyk, J. Bai, and C. R. Hubbard, Surf. Sci. **492**, L711 (2001).
[11] M. R. Castell, Surf. Sci. **505**, 1 (2002).
[12] K. Johnston, M. R. Castell, A. T. Paxton, and M. W. Finnis, Phys. Rev. B **70**, 085415 (2004).
[13] V. Vonk, S. Konings, G. J. van Hummel, S. Harkema, and H. Graafsma, Surf. Sci. **595**, 183 (2005).
[14] F. Silly, D. T. Newell, and M. R. Castell, Surf. Sci. **600**, L219 (2006).
[15] L. M. Liborio, C. G. Sánchez, A. T. Paxton, and M. W. Finnis, J. Phys. Condens. Matter **17**, L223 (2005).
[16] N. Bickel, G. Schmidt, K. Heinz, and K. Müller, Phys. Rev. Lett. **62**, 2009 (1989).
[17] V. Garcia, M. Bibes, A. Barthélémy, M. Bowen, E. Jacquet, J.-P. Contour, and A. Fert, Phys. Rev. B **69**, 052403 (2004).
[18] M. Kawasaki, K. Takahashi, T. Maeda, R. Tsuchiya, M. Shinohara, O. Ishiyama, T. Yonezawa, M. Yoshimoto, and H. Koinuma, Science **266**, 1540 (1994).
[19] G. Koster, B. L. Kropman, G. J. H. M. Rijnders, D. H. A. Blank, and H. Rogalla, Appl. Phys. Lett. **73**, 2920 (1998).
[20] T. Ohnishi, K. Shibuya, M. Lippmaa, D. Kobayashi, H. Kumigashira, M. Oshima, and H. Koinuma, Appl. Phys. Lett. **85**, 272 (2004).
[21] R. Feidenhans'l, Surf. Sci. Rep. **10**, 105 (1989).
[22] I. K. Robinson and D. J. Tweet, Rep. Prog. Phys. **55**, 599 (1992).
[23] P. R. Willmott, C. M. Schlepütz, B. D. Patterson, R. Herger, M. Lange, D. Meister, D. Maden, C. Brönnimann, E. F. Eikenberry, and G. Hülsen et al., Appl. Surf. Sci. **247**, 188 (2005).
[24] C. M. Schlepütz, R. Herger, P. R. Willmott, B. D. Patterson, O. Bunk, C. Brönnimann, B. Henrich, G. Hülsen, and E. F. Eikenberry, Acta Crystallogr. Sect. A **61**, 418 (2005).
[25] Note also, that all evidence of reconstructions vanished after exposure of the surface to less than 1 s (approximately 0.1 monolayer) of flux from hyperthermal particles having kinetic energies of the order of 25 eV during pulsed laser deposition.
[26] O. Bunk, Ph.D. thesis, University of Hamburg, 1999, URL http://www.sub.uni-hamburg.de/opus/volltexte/1999/99/.
[27] Despite this, exchanging the goodness-of-fit criterion from χ^2 to minimization of the crystallographic R factor and back led to the same final model.
[28] W. C. Hamilton, Acta Crystallogr. **18**, 502 (1965).
[29] Q. D. Jiang and J. Zegenhagen, Surf. Sci. **338**, L882 (1995).
[30] O. Warschkow, M. Asta, N. Erdman, K. R. Poeppelmeier, D. E. Ellis, and L. D. Marks, Surf. Sci. **573**, 446 (2004).
[31] R. Herger, P. R. Willmott, O. Bunk, C. M. Schlepütz, B. D. Patterson, and B. Delley (to be published).
[32] B. Delley, J. Chem. Phys. **92**, 508 (1990).
[33] B. Delley, J. Chem. Phys. **113**, 7756 (2000).
[34] J. P. Perdew, K. Burke, and M. Emzerhof, Phys. Rev. Lett. **77**, 3865 (1996).
[35] L. D. Roelofs, G. Y. Hu, and S. C. Ying, Phys. Rev. B **28**, 6369 (1983).
[36] On a speculative note, the disordered (1×1) domains suggested here may, in the cold sample, comprise regions or "grain boundaries" of disorder between the ordered (2×1) and (2×2) reconstructions, frozen in as the sample slowly cools at the end of the substrate preparation procedure.
[37] C. Noguera, J. Phys. Condens. Matter **12**, R367 (2000).
[38] V. Ravikumar, D. Wolf, and V. P. Dravid, Phys. Rev. Lett. **74**, 960 (1995).

Paper II
Surface structure of SrTiO$_3$(001)

The work presented in this chapter has been published in:
R. Herger, P.R. Willmott, O. Bunk, C.M. Schlepütz, B.D. Patterson, B. Delley, V.L. Shneerson, P.F. Lyman, and D.K. Saldin, *Surface structure of SrTiO$_3$(001)*, Phys. Rev. B. **76**, 195435 (2007).

artikel2.pdf is the online version of the library (Zentralbibliothek Zürich).

Abstract

We report on the structural determination of the surface of TiO$_2$-terminated SrTiO$_3$(001) using surface x-ray diffraction. The detailed analysis of two surface diffraction data sets are presented, one (cold) taken at room temperature in vacuum, and the other (hot) under typical conditions used for thin film growth. 49 different combinations of possible surface terminations are described for the cold structure, from which the final structure was chosen, consisting of a weighted mixture of a (1×1) relaxation and (2×1) and (2×2) reconstructions, simultaneously present at the surface. The structures are best modeled by a TiO$_2$-rich surface similar to that proposed by Erdman *et al.* [Nature (London) **419**, 55 (2002).]. The reconstructions are energetically favorable according to density functional theory. They disappear within several minutes upon heating to the hot conditions, forming a termination very similar to the cold (1×1), but more puckered and higher in energy. Six additional models, suggested by direct methods and the literature, to describe the hot surface are also discussed. Direct methods confirm the TiO$_2$-rich termination and the atomic positions of the hot surface. The atomic coordinates for the two TiO$_2$-rich surfaces exhibit significant displacements down to three unit cells, which may have important implications on possible surface ferroelectric phenomena in SrTiO$_3$. Surface energy considerations suggest a temperature-induced order-disorder transition, produced by a mixing of the (2×1) and (2×2) reconstructions, to form the hot pseudo (1×1) structure. Electrostatic stability arguments provide circumstantial support for the experimentally determined TiO_2-rich surfaces.

DOI: 10.1103/PhysRevB.76.195435 [a]

PACS numbers: 61.10.Nz, 68.35.Bs, 68.47.Gh, 71.15.Mb

Reprinted with kind permission from the American Physical Society.

[a] Note that you need a subscription for this journal to directly access the article.

… PHYSICAL REVIEW B **76**, 195435 (2007)

Surface structure of $SrTiO_3(001)$

R. Herger, P. R. Willmott,* O. Bunk, C. M. Schlepütz, and B. D. Patterson
Swiss Light Source, Paul Scherrer Institut, CH-5232 Villigen, Switzerland

B. Delley
Condensed Matter Theory Group, Paul Scherrer Institut, CH-5232 Villigen, Switzerland

V. L. Shneerson, P. F. Lyman, and D. K. Saldin
Department of Physics and the Laboratory for Surface Studies, University of Wisconsin-Milwaukee, Milwaukee, Wisconsin 53201, USA
(Received 6 June 2007; published 21 November 2007)

We report on the structural determination of the surface of TiO_2-terminated $SrTiO_3(001)$ using surface x-ray diffraction. The detailed analysis of two surface diffraction data sets are presented, one (cold) taken at room temperature in vacuum, and the other (hot) under typical conditions used for thin film growth. 49 different combinations of possible surface terminations are described for the cold structure, from which the final structure was chosen, consisting of a weighted mixture of a (1×1) relaxation and (2×1) and (2×2) reconstructions, simultaneously present at the surface. The structures are best modeled by a TiO_2-rich surface similar to that proposed by Erdman *et al.* [Nature (London) **419**, 55 (2002)]. The reconstructions are energetically favorable according to density functional theory. They disappear within several minutes upon heating to the hot conditions, forming a termination very similar to the cold (1×1), but more puckered and higher in energy. Six additional models, suggested by direct methods and the literature, to describe the hot surface are also discussed. Direct methods confirm the TiO_2-rich termination and the atomic positions of the hot surface. The atomic coordinates for the two TiO_2-rich surfaces exhibit significant displacements down to three unit cells, which may have important implications on possible surface ferroelectric phenomena in $SrTiO_3$. Surface energy considerations suggest a temperature-induced order-disorder transition, produced by a mixing of the (2×1) and (2×2) reconstructions, to form the hot pseudo (1×1) structure. Electrostatic stability arguments provide circumstantial support for the experimentally determined TiO_2-rich surfaces.

DOI: 10.1103/PhysRevB.76.195435   PACS number(s): 61.10.Nz, 68.35.Bs, 68.47.Gh, 71.15.Mb

I. INTRODUCTION

Perovskites have been subject of intense research in recent years due to their intriguing physical properties, such as high-temperature superconductivity, colossal magnetoresistance, ferroelectricity, and heterostructures containing two-dimensional electron gases.[1–4] The potential technological applications of perovskites in mass storage devices, as read heads or ferroelectric random access memories, to mention just two examples, underline the importance of understanding the atomic structures of such complex metal oxides. Because in most applications, these materials are used in the form of thin films, it is especially important that their surface and interfacial structures are understood. Subtle structural differences in these regions may lead to fundamentally different physical properties, due to the strong correlation of the valence electrons. On the one hand, surface effects such as reconstructions can set a lower limit to downsizing of devices that exploit bulk effects, while on the other, unexpected new phenomena, for example, ferroelectricity, may occur at the surface.[2,5]

Strontium titanate ($SrTiO_3$, STO) is the paradigmatic substrate material for thin film growth of perovskites. While the surface of STO has been the subject of intense investigations, there remains a notable lack of a detailed description with sub-Å resolution after typical *ex situ* chemical and thermal preparation and, subsequently, under thin film growth conditions. As a result, no concise picture of the surface structure exists, and thus, the STO surface remains a controversial subject. This is partly due on the one hand to the sensitivity of the surface to the preparation and processing conditions, and on the other, to the limited spatial resolution associated with most surface characterization techniques. Consequently, at least nine different surface terminations have been reported for STO(001), depending on the various preparation and ambient conditions, including (1×1), (2×1), (2×2), (6×2), $c(4 \times 2)$, $c(6 \times 2)$, $c(4 \times 4)$, ($\sqrt{5} \times \sqrt{5})R26.6°$, and ($\sqrt{13} \times \sqrt{13})R33.7°$ reconstructions and relaxations.[6–19] Of these observations, two report the simultaneous presence of more than one surface termination at the same time.[9,19]

Of particular relevance among the above terminations to this Report is a Ti-rich (2×1) reconstruction proposed by Erdman *et al.*,[14] where the surface terminates not with one, but two TiO_2 atomic layers (ALs), determined by transmission electron microscopy. The topmost Ti atoms form a characteristic "zigzag" motif, while the Ti sites in the next AL, i.e., one layer deeper down, occupy bulk positions. This TiO_2 double layer (DL) structure, and adaptations of it, have been widely discussed.[12,15,17,19,20]

We recently reported on coexisting DL terminations on STO, consisting of a (2×1) reconstruction similar to that described by Erdman *et al.*,[14] plus a novel (2×2) reconstruction, as well as a TiO_2 (1×1) DL relaxation on the surface of STO.[21] In this work, we present a detailed account of the many models we considered for the large surface x-ray diffraction (SXRD) data set of the STO(001) surface that we acquired, and how we finally determined the DL model de-

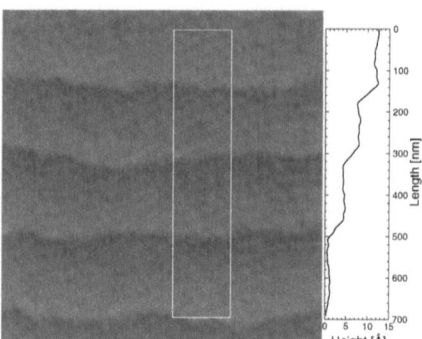

FIG. 1. (Color online) AFM image of STO after chemical etching and thermal annealing. The step height of 4 Å corresponds well to a unit cell step, while no 2 Å steps, indicative of mixed SrO/TiO$_2$ termination, were observed. The line profile (right) was generated by horizontal integration over the region marked by the white box.

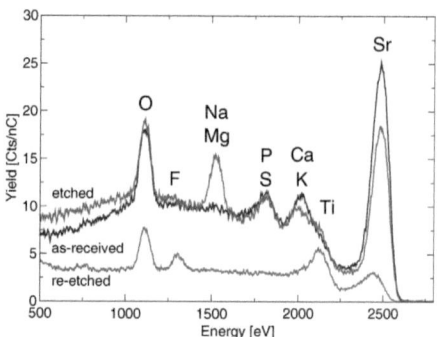

FIG. 2. (Color online) LEIS measurements of as-received (blue) and etched (red) (Refs. 23 and 24) and re-etched STO (green) (Ref. 27). In addition to Sr, Ti, and O, also Ca/K, P/S, Na/Mg, and F are present in the outermost layer. The spectra show decreased intensities for the Sr and Ca species, but an enhanced Ti signal for prepared STO. The main contaminant Mg could be explained by adsorbates arising from the MgO crucible during the anneal process.

scribed in Ref. 21. In arriving this model, we were guided by complementary experimental data, possible models suggested by phase-retrieval methods, and our own theoretical calculations, all of which are also described here. Structures were refined for two data sets, the first taken at room temperature and in ultrahigh vacuum (henceforth referred to as "cold"), and the second recorded at elevated temperature and in an oxygen atmosphere, typical for the growth of perovskite thin films (henceforth referred to as "hot").

II. METHODS

A. Experimental

1. Substrate preparation

STO(001) substrates (space group $Pm\bar{3}m$, $a_0 = 3.9045$ Å) with low vicinality ($<0.1°$) and an impurity content of <5 ppm Ca (Ref. 22) were purchased from CrysTec GmbH, Germany. Substrates of $10 \times 8 \times 1$ mm^3 were prepared according to an established chemical and thermal treatment,[23,24] in order to ensure 100% termination on the TiO$_2$ AL and smooth terrace edge profiles.

2. Characterization

Atomic force microscopy (AFM, Fig. 1) performed on prepared STO revealed a uniform terrace step height of about 4 Å (i.e., one unit cell). Comparison with AFM images of unannealed surfaces (not shown) confirmed that annealing smooths the terrace-edge contours. The terraces are 200 nm wide, consistent with the nominal vicinality.

X-ray photoelectron spectroscopy (XPS) using an Al $K\alpha$ source at 1487 eV was used to confirm TiO$_2$ termination. The angle-dependent electron yield of the oxygen 1s signal showed two peaks that could be assigned to bulk and surface contributions. Due to the insulating properties of STO, only relative shifts in energy, i.e., the chemical shifts, could be established. The chemical shifts between the bulk and surface contributions were 2.6 and 2.0 eV for as-received and prepared STO, respectively. This is consistent with the chemical shifts found for SrO (2.7 eV) and TiO$_2$ (1.9 eV).[25] This suggests a TiO$_2$ termination for etched STO, whereas for unprepared samples both TiO$_2$ and SrO terminations are present.

Low-energy ion scattering (LEIS) was performed by Calipso B.V. at the Technical University of Eindhoven, the Netherlands.[26] The ^{4}He$^+$ ions had an energy of 3 keV. In Fig. 2, we present the LEIS spectra of as-received STO and STO prepared according to Refs. 23, 24, and 27. The red and the blue spectra were taken after exposing the samples to atomic oxygen of thermal energy for 20 min, in order to remove environmental pollution. The green spectrum was taken after an additional etch and rinse step of the etched sample (i.e., of the one that gave rise to the red LEIS spectrum). One or two chemical elements are assigned to each of the peaks, depending on whether the measurement could adequately separate the masses. Section III D will further discuss the LEIS results shown in Fig. 2.

3. Surface x-ray diffraction

SXRD experiments were carried out at the surface diffraction station of the Materials Science Beamline at the Swiss Light Source, Paul Scherrer Institut.[28] The substrates were introduced via a vacuum loadlock into an ultrahigh vacuum (UHV) chamber equipped with a large beryllium window[29] mounted on a (2+3) circle surface diffractometer. Owing to the availability of a fast single photon-counting two-dimensional x-ray pixel detector, it was feasible to reliably record a data set of independent structure factors which was sufficiently large to refine a highly complex structural model within a limited beamtime. From the pixel images, the integrated intensities were extracted and the standard geometrical correction factors applied.[30] Data were recorded using 1 Å synchrotron radiation and a fixed glancing incidence angle of 0.5°, well above the critical angle of 0.20° for total reflection at this energy. Two independent sets of data of two different samples were taken for both the cold and the hot conditions, to confirm reproducibility.

The cold surface showed both crystal truncation rods (CTRs) at integral (h,k) positions in reciprocal space and superstructure rods (SSRs) at half-integral-half-integral positions $[h=m/2, k=n/2; m,n \text{ odd}]$ and half-integral-integral positions $[h=m/2, k=n/2; m \text{ odd/even}, n \text{ even/odd}]$. It became clear during the structure refinement that the latter reflections could not be exclusively associated with a (2×2) reconstruction, and that a (2×1) reconstruction also had to be invoked. Both the CTR and SSR signals were stable under UHV conditions over the measurement time of over 24 h.

The cold data set consists of 9 CTRs and 18 SSRs, comprising a total of 1668 nonequivalent structure factors. In addition, we recorded 806 equivalent data points in order to determine systematic differences between symmetry-equivalent structure factors. Indeed, the total error is primarily associated with such systematic differences, most likely due to (a) the known imperfections of Verneuil-grown STO crystals[31] and (b) a slight bowing of the surface caused by the clamping mechanism to the heater. The error was determined to be 11.7%, and symmetry-equivalent structure factors were averaged. The data set spans $|h|$, $|k|$, and $l=0$ to 3 ($Q \leq 8.36 \text{ Å}^{-1}$).

The SXRD data of the hot sample was taken under typical conditions for thin film growth of perovskites, i.e., the STO was heated to 750±30 °C in 10^{-3} Pa O_2. The SSRs gradually weakened during the heating period of about 30 min., with only CTRs remaining. The SSRs did not return upon subsequent rapid cooling to cold conditions, indicating either an irreversible change in the surface structure or a kinetically hindered transition.

A set of 837 nonequivalent structure factors in 9 CTRs plus 764 equivalent data points was taken under the hot conditions. The systematic error between symmetry-equivalent structure factors was, at 28.9%, significantly higher than for the cold data, and is probably attributable to mechanical distortions produced by thermal strain due to the heater/clamping mechanism. The hot data set also spans reciprocal space for $|h|$, $|k|$, and $l=0$ to 3.

In order to record nonspecular CTRs (and SSRs) in the stationary geometry, i.e., with a detector acceptance that is sufficiently large to integrate the complete diffraction signal over the entire in-plane direction in one single exposure, one can work out the minimum outgoing angle $\beta_{\text{out,min}}$, and (with respect to the STO lattice) the minimum l_{min} of a rod, above which independent structure factor amplitudes for a given sampling resolution Δl are possible.[32] Using an area pixel detector, this limit depends on the quality of the crystal and the energy resolution of the diffractometer. In the case of STO, typical in-plane rocking curves had widths of $\Delta Q_\parallel = 0.0044 \text{ Å}^{-1}$ (i.e., $\Delta \omega = 0.02°$). We thus obtain $l_{\text{min}} = 0.90$ and $l_{\text{min}} = 0.48$ for the sampling resolution used of $\Delta l = 0.025$ and $\Delta l = 0.05$ for CTRs and SSRs, respectively. This means that the diffraction data in the case of the CTRs are not completely resolved in the low-l region. If we assume for this low-l region that only half of the data points are reliable for the CTRs (but all for the SSRs, in agreement with the used sampling resolution), we get 1566 and 727 independent structure factor amplitudes for the cold and the hot data sets, respectively.

B. Theoretical

1. Structure refinement

Both the cold and the hot surface structures were determined using the program FIT.[33] To refine parameters, it uses a robust grid search algorithm. A serious problem in all simple fitting procedures is the possibility of getting trapped in a local minimum. We tried to overcome this obstacle by iterative cycles of resetting fit parameters and refining the structure. For the presented surface structures, typically 50 iterations for both cold and hot STO were needed to obtain a final fit.

Domains were added incoherently. The sharp diffraction signals indicate large domain sizes, while the longitudinal coherence length of the focussed beam is 40 nm at a beam energy of 14 keV for our source.[28] Equivalent domains were equally weighted with regard to the surface symmetries, i.e., no *a priori* preferential orientation of the domains was assumed.

The final goodness-of-fit (GOF) is given in terms of the crystallographic R factor,[34,35] although optimization was carried out via χ^2 minimization, in order to avoid artificial weighting of the most intense signal near the Bragg maxima in the CTRs. To be able to compare different models χ_r^2, i.e., the reduced values,[36] are always given. It is noted that, for the final model, changing from χ^2 to minimization of the R factor and back led to the same final model, within a reasonable set and range of reset fit parameters.

For the structures presented in this work, atomic coordinates, Debye-Waller (DW) factors, a general scaling factor, and, for the cold surface, fractional contributions from each surface-structure type were refined. Atomic movements are therefore represented by up to three parameters per site, depending on the surface symmetry assumed. The underlying bulk atoms had one isotropic DW factor for each Sr, Ti, and O. In each of the lower ALs, isotropic DW factors were used for any particular atom type. Each atom of the top four ALs (or five in the DL surface terminations) was modeled using an anisotropic DW factor, with separate in- and out-of-plane components. As lower limits of the DW factors, 50% of the bulk values for Sr, Ti, and O were chosen. FIT uses DW factors in terms of B values, according to international standards.[37] For convenience, the DW factors presented here are converted into root mean squared displacements (RMSDs), i.e., the uncertainties of the atomic position due to thermal effects are given as the amplitudes of Gaussian-distributed oscillations, represented in units of the STO unit cell (see Tables II–V). Note that the DW factor accounts for both the dynamic motion and the static disorder with Gaussian distribution and may be called the "atomic displacement parameter" to stress explicitly the static component.[38] Unless otherwise stated, every atom had an occupation parameter of unity. Note, however, that a high DW factor may also imply partial occupation. To model the two different surfaces with sets of parameters as described above, we used 394 parameters for the 112 atoms of the cold and 57 for 16 sites of the hot data, implying an oversampling of the independent structure factors by a factor of 4 and more than 10, respectively, i.e., well above the Nyquist criterion.

2. Density functional theory calculations

Density functional (DFT) calculations were performed using periodic slab models with a periodicity normal to the surface of hot STO. This results in a slab separation of about 25 Å for the thickest slabs, and more for thinner slabs. The slabs have a mirror symmetry (possibly combined with glide or inversion symmetry), resulting in a vanishing dipole moment across the slab. DFT equations were solved using the DMOL[3] program, using default settings as described in Refs. 39–41.

The reciprocal space integrations were performed by sampling the Brillouin zone with unshifted 4×4, 2×4, and 2×2 meshes for the (1×1), (2×1), and (2×2) surface unit cells, respectively. To estimate the sensitivity with respect to the density functional approximation, two sets of calculations were carried out, one with a local density functional Perdew-Wang-correlation (PWC),[42] and the other including gradient corrections according to Perdew, Burke, and Ernzerhof (PBE).[43]

Wetting the surfaces by a high dielectric liquid, such as water, may change the relative surface energies. A simple estimate of the wetting effect is obtained by the conductorlike screening model applied to a surface (later referred to as "e-wet"). In this model, dielectric screening charges (i.e., mimmicing a conductorlike charge distribution) are semiempirically applied to the surface in order to simulate the effects of contact with water.[44] Removing the Sr and its complexes by etching is also our rationale for comparing surfaces of different stoichiometry at low Sr chemical potential.

3. Direct methods

The challenge in diffraction based structure determination is the fact that the measured diffraction intensities are proportional to the squared amplitudes of the complex structure factors, but contain no information about their phases (the so-called "phase problem"). Therefore, a direct inversion of the Fourier components to retrieve the electron density in the crystal is impossible.

Conventional fit optimization programs such as FIT and ROD (Refs. 33 and 45) rely on detailed *a priori* knowledge of the system (i.e., the starting model) to overcome this problem. However, the probability of convergence to the correct solution depends very strongly on how well the initial model approximates the true structure. This problem becomes progressively severe with increased system complexity.

Although the missing phase data cannot be measured directly, 2D periodic structures such as crystal surfaces, in contrast to 3D systems (i.e., bulk crystals), may provide this information by other means. Bulk diffraction data do not allow one to measure at least twice as many independent structure factors to reliably determine every single refinement parameter, as demanded by the Nyquist criterion, since only a discrete set of nonzero Fourier amplitudes, the Bragg peaks, can be determined. In surface diffraction data, however, the continuous signal along a CTR provides the necessary redundant information ("oversampling") needed for the application of so-called direct-or phase retrieval methods.

Several direct method attempts to solve the phase problem in surface crystallography have been reported.[46–49] We used the PARADIGM method proposed by Saldin *et al.* to study the surface of hot STO. The algorithm aims to recover the surface structure by an iterative algorithm[50] that alternately satisfies constraints in real and reciprocal space and exploits the fact that scattering from the unknown surface structure may be regarded as a perturbation of that from the truncated bulk structure.

III. RESULTS AND DISCUSSION

A. Room-temperature structure

1. Starting models

The presence of SSRs at half-integral-half-integral positions in k space can in principle be explained by a (2×2) reconstruction completely covering the surface. However, as mentioned above, structural modeling showed that more than one reconstruction was required. A weighted mixture of a (1×1) relaxation and (2×1) and (2×2) reconstructions best reproduced the data.

In the following, we discuss the starting models for the three terminations that were combined to model the cold surface. Coordinates are given in units of the bulk STO cell, with the origin in the lower left corner.

(1) (1×1) *models.* Figure 3 summarizes the cold (1×1) models. The bulk STO unit cell terminates with an AL of TiO_2 (a) placed on top of an AL of SrO (b). For the Ti-rich DL structures, either an AL of TiO_2 (c) or TiO (d) was added on top of the bulk TiO_2 layer (a). The structure in (e) is O deficient and was proposed to be energetically favorable compared to (a).[51] Another rowlike structure that was considered is produced by moving the O in (a) on top of the Ti atom (model f), instead of removing it (as in model e). To test the idea of Sr-enriched surfaces due to segregation during annealing suggested in Ref. 27, we added another SrO AL onto the Ti-rich DL structure shown in (c), either with the Sr in the (1/2 1/2) position (g) or at (0 1/2) (h).

(2) (2×1) *models.* The (2×1) starting models are shown in Fig. 4. The AL in Fig. 4(a) sits on top of a bulk TiO_2 layer, and together they form a TiO_2-rich surface, as proposed by Erdman *et al.*[14] Note that the Ti atoms form a characteristic zigzag motif. Two modifications of the top TiO_2 layer of bulk STO were discussed by Castell[15] in a scanning tunneling microscopy study (STM): (b) is a TiO_2-depleted structure, in which only one TiO_2-unit is present in a (2×1) reconstructed cell, whereas in (c) one O is removed to form a reconstruction with Ti_2O_3 stoichiometry. In (d), the Ti at-

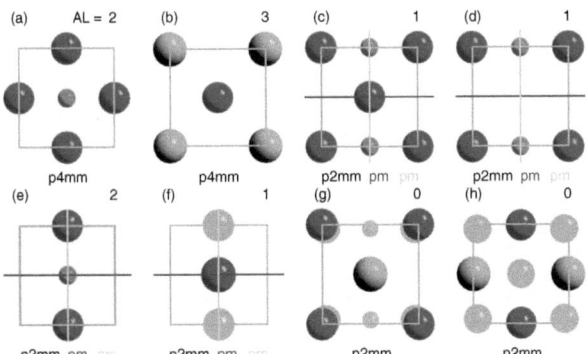

FIG. 3. (Color) Cold (1×1) terminations that are placed on STO bulk unit cells. The first two ALs of an ideal Ti-terminated bulk STO surface are shown in (a) and (b). All models are viewed from above onto the *ab* plane and display the unit cell as a grey box, with the Sr bulk position at the cell corners. The number in the upper right corner indicates the AL of the opaque sites, labeling the surface layer of DL models as 1. The adopted surface symmetries are noted in black below the cells where they are unambiguous. Other considered symmetry elements and their descriptions are highlighted in brown and yellow. Atoms shown in lighter shades represent nominal positions that are one AL deeper down, but are required to unambiguously understand the termination. A detailed description of the structures is given in the text. Color code: Sr green, Ti red, and O blue.

oms also form a zigzag motif, but in contrast to (a), this is a modification of the top TiO_2 layer of bulk STO and is not a DL structure. The models in (e) and (f) are again O deficient, forming, respectively, a rowlike structure (as suggested by Noguera[51]), or a structure in which an O atom is moved on top of the adjacent Ti atom, similar to the cold (1×1) structure in Fig. 3(f). Assuming that an O atom moves with equal probability from the bulk position to the left or the right Ti atom, the occupation of the top O was set to 0.5 in model (f). Model (g) contains a Sr adatom layer with one Sr atom per (2×1) cell, as proposed in a STM study by Kubo and Nozoye.[12] We note that the stability of Sr adatom surfaces has been questioned for high Sr coverages and could only be explained by first-principle calculations if the surface is far from equilibrium.[20] Finally, models (h) and (i) have the same basic idea as presented in Figs. 3(g) and 3(h), respectively, i.e., an additional SrO layer on a Ti-rich DL structure.

(3) (2×2) models. The (2×2) reconstructions are shown in Fig. 5. Structures (a)–(c) all represent Ti-rich DLs on bulk TiO_2 layers. In model (a), the fully stoichiometric TiO_2 top layer consists of the characteristic zigzag motif and was found to be energetically favorable by DFT calculations presented by Warschkow et al.[52] Note that this reconstruction is very similar to the (2×1) structure in Fig. 4(a): it can be

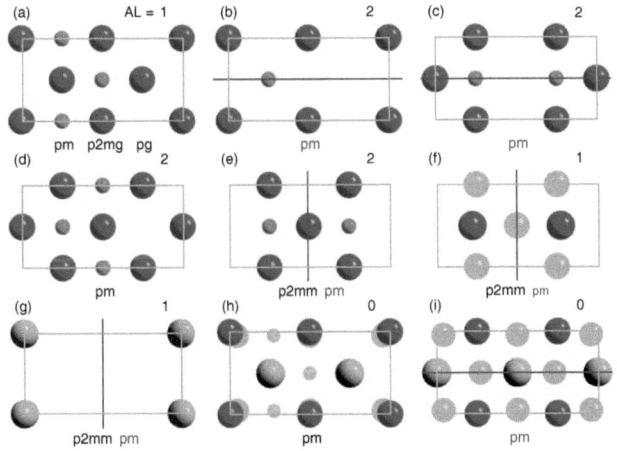

FIG. 4. (Color) The tested starting models for the cold (2×1) structures, using the same representation as in Fig. 3.

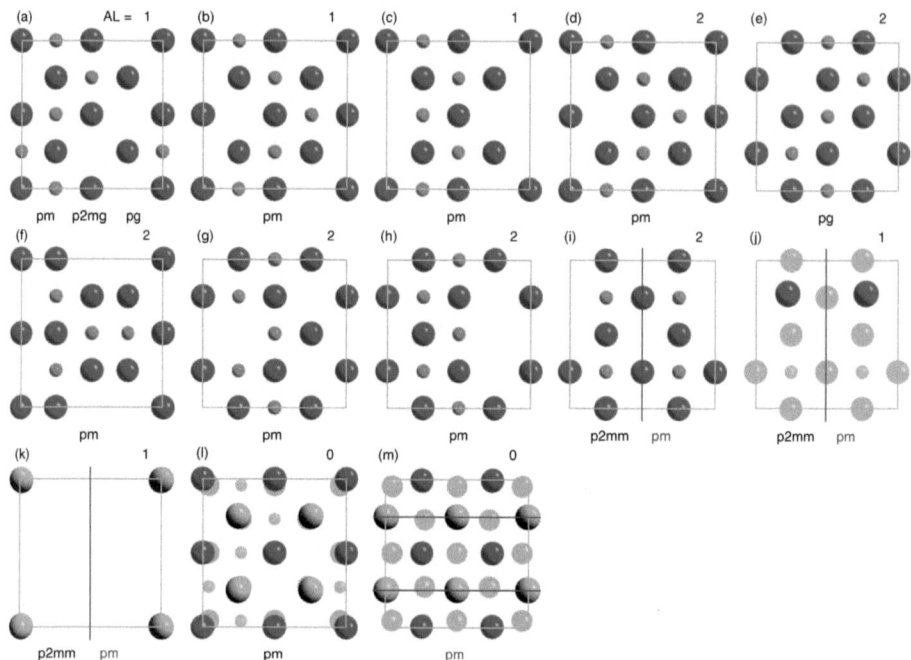

FIG. 5. (Color) Starting models for the cold (2×2) structures. Notation is the same as for Fig. 3.

thought of as consisting of a juxtaposition of such a (2×1) structure and its mirror image. Structure (b) is very similar to (a), whereby the Ti atom zigzag motif is shifted by (1/2 1/2) of a unit cell. Structure (c) consists of two adjacent (2×1) reconstructions of the type shown in Fig. 4(a), but due to O deficiency at the (0 1) position, it forms a true (2×2) reconstruction. The structures (d)–(h) are also based on Ti atom rearrangements, but in contrast to (a)–(c), no double Ti-layer is present, i.e., the changes take place in the top TiO_2 layer of bulk STO. Model (d) looks thus the same as (b). The motif in (e) is similar to that in (a), but with the important differences of having two Ti sites on bulk Sr positions and the other two Ti atoms directly above bulk Ti. Model (f) shows an unusual Ti and O arrangement that would be difficult to explain by electrostatic arguments, but is presented for the sake of completeness. The two structures (g) and (h) are also based on the idea of two adjacent (2×1) reconstructions (with the Ti on top of bulk Sr/Ti again) and O deficiencies [missing either the (1/2 1) or the (3/2 1) site] forming the (2×2) structures. For the models (i) and (j), we were inspired by an O deficiency (i) or an O movement on top of a bulk Ti site (j), analogous to structures already presented for the (1×1) and (2×1) models. The same is true for the Sr adatom model[12] (k), or SrO overlayers on a TiO_2-rich DL structure to form the structures (l) and (m), model (l) having varying Sr and O occupation (see Table I).

2. Refinement

The models presented in Sec. III A 1 were combined and fit to the experimental data. Table I shows the most promising 49 combinations. We note that, in assessing the potential of a model, we not only considered the χ_r^2 or crystallographic R factor, but also its physical reasonableness. These included sensible atom positions and distances between different ALs, or plausible DW factors. The fit procedure used a GOF criterion that was a weighted combination of the χ_r^2 and a punitive energy term proportional to the square of the deviation from equilibrium bond lengths, similar to the "Keating energy" used in covalent structures.[53] This second term helped to stabilize the models during the first few iterations. In general, we started to fit the top layers only, then subsequently added deeper ALs as the model relaxed. The weighting of the punitive energy term could then be reduced and the refinement iterated.

In arriving at the final model (number 36 in Table I), we were guided by a clear and consistent trend: all models containing a TiO_2 DL and a zigzag motif of the top Ti atoms produced significantly better fits than those without these features. In addition, these models had sensible atomic positions and DW factors throughout.

The (2×1) and (2×2) reconstructions in models 1–6 have a higher symmetry than other models in Table I. The resulting GOF values (χ_r^2 and R factor) are high compared to

TABLE I. Selected surface structure combinations under identical fit conditions. The letters in column (1×1), (2×1), and (2×2) correspond to those shown in Figs. 3–5, respectively. Subscripted h and v depict horizontal and vertical mirror planes, respectively. Column AL gives the number of fit atomic layers, the last three columns are the fractional contributions of the terminations.

Model	(1×1)	(2×1)	(2×2)	Surface symmetries			ALs	$\chi_r^2(|F|^2)$	$R(|F|)$	$f(1\times1)$	$f(2\times1)$	$f(2\times2)$
1	a	e	i	p4mm	p2mm	p2mm	4	4.416	0.154	0.150	0.316	0.534
2	a	f	j	p4mm	p2mm	p2mm	4	6.218	0.166	0.188	0.307	0.505
3	f	f	j	p2mm	p2mm	p2mm	4	5.686	0.164	0.108	0.312	0.580
4	a	g	k	p4mm	p2mm	p2mm	5	3.745	0.140	0.341	0.328	0.331
5	c		a	p2mm		p2mg	5	5.146	0.146	0.437		0.563
6	c	a	a	p2mm	p2mg	p2mg	5	2.903	0.131	0.361	0.180	0.459
7	a	f	j	p4mm	pm_v	pm_v	4	2.556	0.144	0.214	0.359	0.427
8	f	f	j	pm_v	pm_v	pm_v	4	2.647	0.135	0.167	0.411	0.422
9	a	e	i	p4mm	pm_v	pm_v	4	3.025	0.140	0.135	0.377	0.488
10	a	b	e	p4mm	pm_h	pg	4	3.058	0.134	0.000	0.501	0.499
11	a	b	d	p4mm	pm_h	pm_h	4	2.228	0.133	0.238	0.340	0.422
12	a	b	f	p4mm	pm_h	pm_h	4	2.313	0.127	0.172	0.352	0.476
13	a	b	b	p4mm	pm_h	pm_h	4	2.282	0.131	0.203	0.329	0.468
14	a	c	e	p4mm	pm_h	pg	4	2.571	0.132	0.214	0.418	0.368
15	a	c	d	p4mm	pm_h	pm_h	4	2.287	0.132	0.145	0.391	0.464
16	a	c	f	p4mm	pm_h	pm_h	4	2.759	0.137	0.000	0.468	0.532
17	a	c	b	p4mm	pm_h	pm_h	4	2.328	0.128	0.145	0.416	0.439
18	a	a	e	p4mm	pm_h	pg	4	2.909	0.123	0.169	0.447	0.384
19	a	a	d	p4mm	pm_h	pm_h	4	2.169	0.122	0.252	0.470	0.278
20	a	a	f	p4mm	pm_h	pm_h	4	2.351	0.115	0.187	0.468	0.345
21	a	d	g	p4mm	pm_h	pm_h	4	2.129	0.111	0.297	0.380	0.323
22	a	d	h	p4mm	pm_h	pm_h	4	2.183	0.118	0.256	0.341	0.403
23	a		b	p4mm		pm_h	5	3.468	0.135	0.385		0.615
24	a	g	k	p4mm	pm_v	pm_v	5	2.624	0.132	0.263	0.345	0.392
25	a	a	b	p4mm	pm_h	pm_h	5	2.069	0.118	0.185	0.386	0.429
26		a	b		pm_h	pm_h	5	2.211	0.120		0.516	0.484
27	c	a	b	p2mm	pm_h	pm_h	5	1.978	0.114	0.218	0.422	0.360
28	c	a	b	pm_h	pm_h	pm_h	5	2.059	0.113	0.225	0.383	0.392
29	c	a	b	pm_v	pm_h	pm_h	5	1.869	0.113	0.208	0.421	0.371
30	d	a	b	p2mm	pm_h	pm_h	5	2.172	0.119	0.170	0.422	0.408
31	d	a	b	pm_h	pm_h	pm_h	5	2.190	0.113	0.173	0.378	0.449
32	d	a	b	pm_v	pm_h	pm_h	5	1.878	0.112	0.217	0.431	0.352
33	c	a^a	c	p2mm	pm_h	pm_h	5	1.771	0.114	0.272	0.341	0.387
34	c	a^a	c	p2mm	pm_h	pm_h	5	2.059	0.108	0.187	0.379	0.434
35	c	a	c	p2mm	pm_h	pm_h	5	1.713	0.116	0.298	0.353	0.349
36	c	a	a	p2mm	pm_h	pm_h	5	1.913	0.118	0.244	0.363	0.393
37	c	a	a	p2mm	pm_h	p2mg	5	2.859	0.116	0.215	0.443	0.342
38	c	a	a	p2mm	pm_h	pg	5	2.263	0.120	0.187	0.419	0.394
39	c	a^b	a^b	p2mm	$p2mg/pm_h$	$p2mg/pm_h$	5	2.154	0.118	0.251	0.366	0.383
40[c]	c	a^b	a^b	p2mm	pg/pm_h	pg/pm_h	5	1.784	0.116	0.190	0.401	0.409
41[c]	c	a	a^b	p2mm	pm_h	$p2mg/pm_h$	5	1.873	0.114	0.228	0.343	0.429
42[c]	c	a	a^b	p2mm	pg/pm_h	pg/pm_h	5	1.580	0.112	0.224	0.323	0.453
43[c]	c	a^a	a^b	p2mm	pm_h	pg/pm_h	5	1.676	0.120	0.168	0.404	0.428
44[c]	c	a^b	a^b	p2mm	pm_h	$pg/p2mg$	5	2.551	0.118	0.218	0.405	0.377
45[c]	g^d	h^d	l^d	p2mm	pm_h	pm_h	6	1.506	0.105	0.342	0.352	0.306
46[c]	g^d	h^d	l^d	p2mm	pm_h	pm_h	5	1.393	0.102	0.181	0.265	0.554
47[c]	g^e	h^d	l^d	p2mm	pm_h	pm_h	6	1.265	0.103	0.292	0.299	0.409
48[c]	g^e	h^e	l^e	p2mm	pm_h	pm_h	5	1.335	0.105	0.238	0.350	0.412
49[c]	h	i	m	p2mm	pm_h	pm_h	6	1.971	0.136	0.000	0.427	0.573

[a]O moved out of the plane, similar to Ref. 14.
[b]Top layer has other symmetry than underneath.
[c]Refined atomic positions unphysical.
[d]Sr and O occupation=1.0.
[e]Sr and O occupation=0.5.

other models. This suggests that reconstructions with lower surface symmetries are preferred.

Models 7–25 all have a TiO_2-terminated (1×1) structure without a double layer. All the χ_r^2 values are higher than 2.1. Note that models with a Ti atom zigzag motif often display low χ_r^2 and R values within this group.

Models 27–44 contain Ti-rich DLs. The χ_r^2 values are around 2 with R factors of 0.120 or lower. Also, the fractional contributions of the different terminations [about 20% (1×1), 40% (2×1), and 40% (2×2)] vary less from model to model than for models 7–25. The first set (27–36) are modeled with $p2mm$ and pm symmetries, whereas in the second set (37–44) glide mirror symmetries ($p2mg$ and pg) are also used, as these belong naturally to the zigzag motif. If the glide symmetry acted on the entirety of the (2×2) domain, the fit was poorer and the models (37 and 38) were physically questionable. By introducing a physically dubious mixed-symmetry within either the (2×1) or (2×2) reconstructions (i.e., the top layer having glide symmetry and everything below, a mirror plane, models 39–43) the fits improved considerably. However, as can be seen in Table I, these models have been deemed to be unphysical. This somewhat subjective judgement was based on the fact that the majority of the atomic positions are unphysically disordered, although the shifts of the individual atoms are generally not unreasonable.

In contrast, our preferred model 36, contains atom positions that appear to remain closer to the starting positions, especially below the top two or three AL.[21] Although it does not have the lowest χ_r^2 at this point of the refinement procedure, it was chosen for the following reasons. First, the model is supported by previous experimental evidence regarding the DL (2×1) domain,[14] and also by DFT calculations, which predict the low surface energy of STO terminated with the DL (2×2) described here.[52] Indeed, our own DFT calculations (see Sec. III C) have shown that the three lowest surface energy configurations include the (2×1) and (2×2) reconstructions of model 36. The atomic positions as well as the DW factors (for the surface and the bulk) seemed to be most physically reasonable, compared to all other DL structures. In particular, the Ti atoms of the zigzag motif, which seems to be so important for producing a low surface energy, barely shift from their start positions.

The GOF of combinations containing a DL-(1×1) relaxation are better than those with a bulk (1×1) termination, but are nevertheless of comparable magnitude. At the point of the refinement procedure presented in Table I, adopting a TiO_2-DL for the (1×1) termination does therefore not seem to be necessary. There are, however, pressing arguments to postulate such a DL structure. First, the combination of DL (2×1) and (2×2) reconstructions with a bulk (1×1) domain would imply half unit cell (i.e., 2 Å) height steps at their boundaries, of which we see no evidence in the AFM pictures (see Fig. 1). A second argument lies in the fact that the reconstructions vanish upon annealing. It seems unlikely that a mostly DL-reconstructed surface transforms to a bulk-like terminated (1×1) when the substrate is brought to conditions typical for thin film growth, which would require the evaporation of the uppermost TiO_2 AL of the (2×1) and (2×2) domains, despite the fact that DFT calculations predict that they should be the most stable configurations. Moreover, the existence of an "ideal" bulk (1×1) termination in UHV has been recently questioned,[17] based on arguments of surface oxygen loss. Finally, the TiO_2-DL (1×1) model fits the experimental data better than a bulk (1×1), when the number of fitted AL is increased.

Models 45–49 investigate the possibility of SrO segregation during the annealing process.[27] The models consisting of such terminations (i.e., models 45–48), have χ_r^2 values and R factors that are clearly superior than those of any of the other presented models. This is, however, at the expense of physical reasonableness. The fitted structures all show the same trend: The top SrO layer is repelled significantly towards vacuum for all terminations, resulting in an interlayer distance of over 3 Å, instead of 2 Å. In addition, the side views of the structures appear to be considerably bent.

3. Final structure

The experimental data and the calculated intensities of model 36 are shown in Fig. 6. The intensity in the SSRs varies strongly with a periodicity of the order of 1/3 of a reciprocal lattice unit. Thus, significant atomic displacements are expected down to three unit cells, which, indeed, we found in the refinement. The quality of the data and the fits is evidenced by a final R factor of 4.5% and the absence of unphysical positions or DW factors. The structure of the final fit model is shown in Figs. 7(a)–7(c).[54] The reconstructions dominate the surface termination with percentages of 43% of (2×2) (c) and 37% of (2×1) (b). A smaller contribution of 20% (a) could be attributed to a (1×1) relaxation. Analysis of four rods of a second STO substrate gave very similar intensities (within 5%) and thus essentially identical weightings of the different reconstruction/relaxation domains. The content of impurities may play a role as well, as argued in Sec. III D.

The atomic coordinates of the (1×1), (2×1), and (2×2) domains are listed in Tables II–IV, respectively. The atoms are grouped in ALs with a positive z axis pointing out of the surface, and the $z=0$ position referring to that layer where the atoms are not shifted, i.e., they are on bulk positions. The first and the second ALs form the DL structure. The fourth column represents the high-symmetry starting position of the atom. The coordinates are given in fractions of bulk STO unit cells, with the associated DW factors as RMSDs. The atoms near the surface generally undergo more pronounced movements than do atoms deeper down and have larger DW factors. A large DW factor may reflect less than unit occupation for some of the atoms, although we have no evidence of vacancies from our fits. Note that for the (1×1) relaxation, no in-plane movements are allowed due to the $p2mm$ symmetry. A further discussion of the cold structure with regards to an order-disorder transition will be given in Sec. III E.

B. Hot structure

1. Starting models

The hot data set consisted of CTRs, with no SSRs present. Thus, we only considered the set of (1×1) terminations

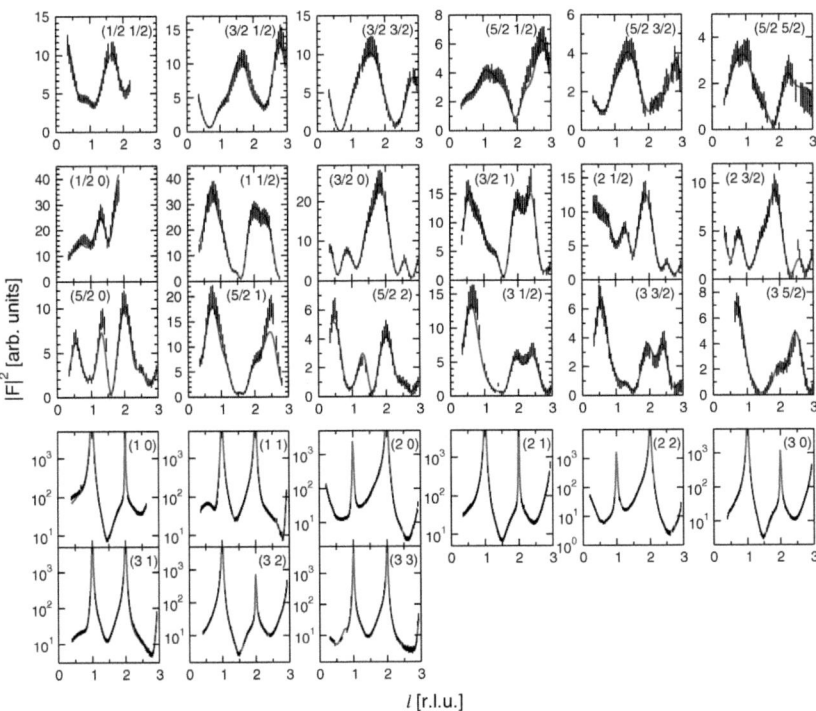

FIG. 6. (Color online) Set of the SXRD data (black data points) and calculated intensities (gray solid line, red online) for the cold conditions. The top row contains the six SSRs that can be exclusively associated with the (2×2) reconstruction. The central group of 12 SSRs can represent signal from both the (2×2) and (2×1) domains, and indeed we see that their intensities are about double those of the uppermost row, indicative that a (2×1) does exist. The nine CTRs at integer positions in k space are presented on a logarithmic scale in the lowest group.

shown in Fig. 8. All these (1×1) structures contain double layers.

Figure 8(a) shows the top AL of the TiO$_2$ DL structure that was also found in the cold data. The oxygen overlayer structure in (b) was recently suggested by Vonk et al.,[18] where they studied a chemically etched and thermally treated (1×1) STO(001) surface via SXRD at room temperature and in air.

As for the reconstructed (cold) case, both p4mm symmetry and multiple domains of p2mm are consistent with the diffraction pattern; PARADIGM treats these cases separately. When p2mm symmetry was assumed, PARADIGM suggested the same TiO$_2$-DL surface as for the cold model (model a). This result will be discussed further in Sec. III B 3. When p4mm symmetry was imposed, PARADIGM suggested four further starting models, all of which contain a metal-rich top surface layer, shown in models (c)–(f). The top AL either consists of Sr, Ti or their mixture. Note that all surfaces suggested by PARADIGM contain a double layer. All of these models were subsequently tested using conventional structure refinement with fit, as described below.

2. Refinement

The different structural models for the hot data set were fit using identical refinement procedures. In Fig. 9, we present the evolution of the GOF criteria χ_r^2 (a) and R factor (b) as a function of the number of fit ALs. All the models were directly fit down to the shown number of ALs. The models are labeled corresponding to Fig. 8, plus a bulk model without a Ti-rich DL.

The best results are obtained when the TiO$_2$-DL structure is fit to the data, both in terms of χ_r^2 and the R factor. The bulk and the O overlayer fit worse than the DL structure. In particular, the O overlayer model reaches unphysical positions for 3 and 4 ALs, i.e., the O atoms are compressed into the next underlying layer. Indeed, the shapes of the CTRs differ significantly between the SXRD data presented here and that from Vonk et al.,[18] which is perhaps not surprising,

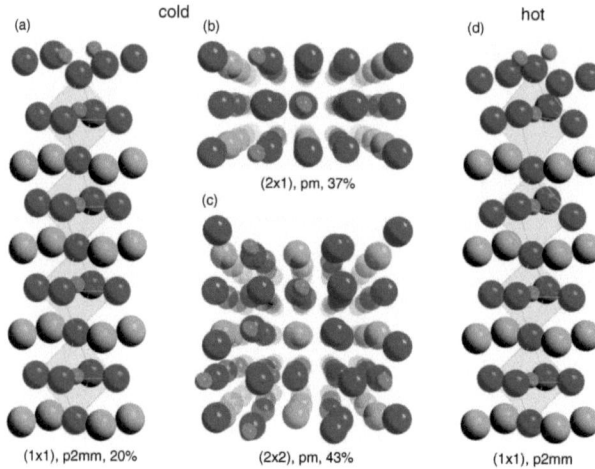

FIG. 7. (Color) The final models for the cold domains (a)–(c) and the hot structure (d), including their symmetries and percentage contributions. The (1×1) structures [cold (a) and hot (d)] are viewed from the side, while the reconstructions are from above. Color code: Sr green, Ti red, and O blue.

given that their study was performed in air and at room temperature, whereas our data was taken under film growth conditions. We also note that we have observed several times that metal oxides samples in air exposed to glancing-incidence synchrotron radiation show clear radiation damage within less than an hour. Also, from a chemical perspective, a dense packing of the surface with O is difficult to explain, either with hydrogen-bonded oxygen, where the experimentally found bonds are too short, or with covalent bonds, where they are too long. Moreover, it is rather unlikely for a perovskite system to have mostly ionic bonds in the bulk and a covalent nature at the surface.

The bulk model consistently produces marginally poorer results than the TiO_2-DL model. Both models give similar results when only 1 AL was fit, but successive layers improve the quality of the fit for the DL structure compared to bulk (see Fig. 9). Moreover, as already argued in Sec. III A 2, a pure bulk termination evolving upon heating from the cold DL structure seems physically improbable.

The metal-rich surfaces are significantly worse than the models discussed so far, the only exception being model (f), a pure Ti surface layer. The other three terminations result in much higher χ_r^2 values and R factors, in particular for the fits of the top 1 or 2 ALs, where improvements in the fit, assuming the model to be correct, should have the most influence. It can be perhaps argued that the reason that the Ti model (f) fits the experimental data a little better is because the TiO_2-DL structure is also Ti-rich at the surface. Further investigations compared the two structures and gave consistently better results for the TiO_2-DL model. Both models looked physically reasonable and had sensible DW factors, but the χ_r^2 was always about 15% better for the TiO_2-DL model, and any attempt to reach the same χ_r^2 with model (f) failed. Another argument supporting the TiO_2-DL model is that the cold structure did not exhibit any sign of a metal-rich surface. To obtain a metal-rich surface would therefore require comprehensive changes of the surface configuration upon annealing in a background of oxygen, which appears to be physically implausible.

3. PARADIGM results

The application of the PARADIGM algorithm on the hot data set recovers electron densities (ED) in an additional layer on top of the STO bulk unit cell, as shown in Fig. 10. The EDs are located at exactly the same positions as one would expect it for the TiO_2-DL model, although with higher

TABLE II. Refined positions and DW factors (expressed as root mean squared displacements in brackets) for the (1×1) reconstruction under cold conditions in units of bulk STO.

AL	No.	Atom	Nominal position	x/a_0	y/a_0	z/a_0
1	1	Ti	$(\frac{1}{2}\,0\,3)$	0.500(54)	0.000(54)	3.187(69)
	2	O	$(0\,0\,3)$	0.000(20)	0.000(20)	3.107(20)
	3	O	$(\frac{1}{2}\,\frac{1}{2}\,3)$	0.500(118)	0.500(118)	2.940(57)
2	4	Ti	$(\frac{1}{2}\,\frac{1}{2}\,\frac{5}{2})$	0.500(14)	0.500(14)	2.529(34)
	5	O	$(\frac{1}{2}\,0\,\frac{5}{2})$	0.500(25)	0.000(25)	2.457(204)
	6	O	$(0\,\frac{1}{2}\,\frac{5}{2})$	0.000(20)	0.500(20)	2.442(20)
3	7	Sr	$(0\,0\,2)$	0.000(55)	0.000(55)	1.916(132)
	8	O	$(\frac{1}{2}\,\frac{1}{2}\,2)$	0.500(20)	0.500(20)	1.962(64)
4	9	Ti	$(\frac{1}{2}\,\frac{1}{2}\,\frac{3}{2})$	0.500(127)	0.500(127)	1.455(72)
	10	O	$(\frac{1}{2}\,0\,\frac{3}{2})$	0.500(20)	0.000(20)	1.462(20)
	11	O	$(0\,\frac{1}{2}\,\frac{3}{2})$	0.000(20)	0.500(20)	1.458(20)
5	12	Sr	$(0\,0\,1)$	0.000(68)	0.000(68)	0.975(197)
	13	O	$(\frac{1}{2}\,\frac{1}{2}\,1)$	0.500(78)	0.500(78)	0.964(61)
6	14	Ti	$(\frac{1}{2}\,\frac{1}{2}\,\frac{1}{2})$	0.500(44)	0.500(44)	0.546(44)
	15	O	$(\frac{1}{2}\,0\,\frac{1}{2})$	0.500(20)	0.000(20)	0.505(20)
	16	O	$(0\,\frac{1}{2}\,\frac{1}{2})$	0.000(20)	0.500(20)	0.492(20)

TABLE III. Refined positions and DW factors (expressed as root mean squared displacements in brackets) for the (2×1) reconstruction under cold conditions in units of bulk STO.

AL	No.	Atom	Nominal position	x/a_0	y/a_0	z/a_0
1	1	Ti	$(\frac{1}{2}\,0\,3)$	0.481(59)	0.000(59)	3.194(16)
	2	O	$(0\,0\,3)$	−0.005(20)	0.000(20)	3.119(132)
	3	O	$(1\,0\,3)$	1.032(20)	0.000(20)	2.975(20)
	4	Ti	$(1\,\frac{1}{2}\,3)$	0.972(204)	0.500(204)	3.345(19)
	5	O	$(\frac{1}{2}\,\frac{1}{2}\,3)$	0.584(57)	0.500(57)	3.260(40)
	6	O	$(\frac{3}{2}\,\frac{1}{2}\,3)$	1.500(44)	0.500(44)	3.266(20)
2	7	Ti	$(\frac{1}{2}\,\frac{1}{2}\,\frac{5}{2})$	0.536(56)	0.500(56)	2.564(14)
	8	O	$(\frac{1}{2}\,0\,\frac{5}{2})$	0.575(20)	0.000(20)	2.637(20)
	9	O	$(0\,\frac{1}{2}\,\frac{5}{2})$	0.017(20)	0.500(20)	2.627(20)
	10	Ti	$(\frac{3}{2}\,\frac{1}{2}\,\frac{5}{2})$	1.590(67)	0.500(67)	2.647(70)
	11	O	$(\frac{3}{2}\,0\,\frac{5}{2})$	1.485(20)	0.000(20)	2.570(20)
	12	O	$(1\,\frac{1}{2}\,\frac{5}{2})$	0.964(109)	0.500(109)	2.579(20)
3	13	Sr	$(0\,0\,2)$	0.016(34)	0.000(34)	2.020(58)
	14	O	$(\frac{1}{2}\,\frac{1}{2}\,2)$	0.600(20)	0.500(20)	2.057(20)
	15	Sr	$(1\,0\,2)$	1.026(39)	0.000(39)	2.018(51)
	16	O	$(\frac{3}{2}\,\frac{1}{2}\,2)$	1.512(204)	0.500(204)	2.031(204)
4	17	Ti	$(\frac{1}{2}\,\frac{1}{2}\,\frac{3}{2})$	0.511(30)	0.500(30)	1.560(23)
	18	O	$(\frac{1}{2}\,0\,\frac{3}{2})$	0.521(20)	0.000(20)	1.515(20)
	19	O	$(0\,\frac{1}{2}\,\frac{3}{2})$	0.025(20)	0.500(20)	1.491(39)
	20	Ti	$(\frac{3}{2}\,\frac{1}{2}\,\frac{3}{2})$	1.516(28)	0.500(28)	1.535(14)
	21	O	$(\frac{3}{2}\,0\,\frac{3}{2})$	1.484(20)	0.000(20)	1.425(20)
	22	O	$(1\,\frac{1}{2}\,\frac{3}{2})$	1.009(29)	0.500(29)	1.490(86)
5	23	Sr	$(0\,0\,1)$	0.015(25)	0.000(25)	1.003(45)
	24	O	$(\frac{1}{2}\,\frac{1}{2}\,1)$	0.499(88)	0.500(88)	1.008(140)
	25	Sr	$(1\,0\,1)$	1.010(25)	0.000(25)	1.007(37)
	26	O	$(\frac{3}{2}\,\frac{1}{2}\,1)$	1.606(23)	0.500(23)	1.061(20)
6	27	Ti	$(\frac{1}{2}\,\frac{1}{2}\,\frac{1}{2})$	0.489(27)	0.500(27)	0.506(27)
	28	O	$(\frac{1}{2}\,0\,\frac{1}{2})$	0.547(50)	0.000(50)	0.499(50)
	29	O	$(0\,\frac{1}{2}\,\frac{1}{2})$	−0.007(50)	0.500(50)	0.498(50)
	30	Ti	$(\frac{3}{2}\,\frac{1}{2}\,\frac{1}{2})$	1.498(27)	0.500(27)	0.504(27)
	31	O	$(\frac{3}{2}\,0\,\frac{1}{2})$	1.518(50)	0.000(50)	0.529(50)
	32	O	$(1\,\frac{1}{2}\,\frac{1}{2})$	1.020(50)	0.500(50)	0.525(50)

intensities on O than on the Ti positions. The agreement in the heights of recovered ED by PARADIGM (see Fig. 10) and the FIT results (see Table V) is striking. Therefore, these results strongly support the double layer model.

The data in Fig. 10 represents the averaged ED for two coherently added domains.[55] The ratio of the recovered ED maxima between (a) and (b) is approximately 3:2. A possible explanation could involve the presence of Sr impurities on the surface, as it is indicated by our LEIS data (see Fig. 2). However, our attempts to model such impurities with FIT resulted in less than 1.5% Sr remaining at the surface, clearly insufficient to account for this inversion of ED as seen by PARADIGM. The origin of the observed EDs therefore remains unclear.

PARADIGM may be applied to SXRD data from either coherently or incoherently scattering symmetrically related domains.[56–58] There is a probability that the present data arise from a mixture of coherently and incoherently scattering domains. The intensities of diffraction rods with $(h+k)$ even would be expected to be the same whether the domains scatter coherently or incoherently, and are therefore the most appropriate to be used as input to the algorithm in this case. On the other hand, the contributions to the intensities of these rods of even $(h+k)$ from the two domains of $p2mm$ symmetry rotated relative to each other by 90° is the same. The Fourier transform of surface structure factors of even $(h+k)$ only will necessarily have $p4mm$ symmetry and may be identified with the average electron density of the two domains that is seen in Fig. 10.

The only contradiction between the surface electron density maps from PARADIGM seen in Fig. 10 and the best-fit model suggested by FIT is that the latter suggests the heavier atom (Ti) to be associated with the smaller electron density features in the figure and that the lighter atom (O) with the higher density. It is entirely possible that this apparent reversal of electron densities is due to the fact that an ideal substrate is assumed in the PARADIGM calculations, whereas FIT suggests significant reconstructions below the outermost double layer. Nevertheless, the model-independent nature of the PARADIGM result gives greater confidence to the conclusions of the trial-and-error fitting to the data of specific structural models.

4. Final structure

The SXRD data of STO under hot conditions is shown in Fig. 11. The best final surface structure has a crystallographic R factor of 11.2% and no unphysical parameters. The structure is shown from the side in Fig. 7(d).[54] It is very similar to the DL (1×1) structure for the cold data, but with more puckering of the TiO_2 AL, as the topmost Ti atom moves further out of the surface.

In Table V, the atomic coordinates are presented in the same manner as for the cold data. The uncertainty in the positions due to thermal effects is higher compared to the cold data, as one would expect. The displacements in the z direction from high-symmetry positions in the starting models are further discussed in Sec. III F 1.

C. DFT results

The DFT results are summarized in Table VI. Seven models were tested, including the most promising DL structures. A comparison between our own calculation and recent work by other groups was also included.[17,52,59]

We focus first on the surface energies that were calculated using the PBE functional. The three lowest surface energies included the two reconstructions we found experimentally. As already mentioned, the DL (2×2) surface can be thought

TABLE IV. Refined positions and DW factors (expressed as root mean squared displacements in brackets) for the (2×2) reconstruction under cold conditions in units of bulk STO.

AL	No.	Atom	Nominal position	x/a_0	y/a_0	z/a_0	AL	No.	Atom	Nominal position	x/a_0	y/a_0	z/a_0
1	1	Ti	$(\frac{1}{2} 0 3)$	0.413(160)	−0.055(160)	2.920(98)	4	33	Ti	$(\frac{1}{2} \frac{1}{2} \frac{3}{2})$	0.525(59)	0.500(59)	1.478(152)
	2	O	(0 0 3)	0.062(20)	0.112(20)	3.015(93)		34	O	$(\frac{1}{2} 0 \frac{3}{2})$	0.472(59)	−0.013(59)	1.487(204)
	3	O	(1 0 3)	1.378(67)	−0.037(67)	2.798(20)		35	O	$(0 \frac{1}{2} \frac{3}{2})$	−0.061(40)	0.500(40)	1.398(33)
	4	Ti	$(0 \frac{1}{2} 3)$	−0.085(73)	0.500(73)	3.013(14)		36	Ti	$(\frac{3}{2} \frac{1}{2} \frac{3}{2})$	1.484(65)	0.500(65)	1.595(14)
	5	O	$(\frac{3}{2} \frac{1}{2} 3)$	1.570(20)	0.500(20)	3.001(20)		37	O	$(\frac{3}{2} 0 \frac{3}{2})$	1.415(67)	−0.005(67)	1.482(20)
	6	O	$(\frac{1}{2} \frac{1}{2} 3)$	0.534(29)	0.500(29)	2.885(20)		38	O	$(1 \frac{1}{2} \frac{3}{2})$	0.921(51)	0.500(51)	1.466(20)
	7	Ti	$(\frac{1}{2} 1 3)$	0.413(160)	1.055(160)	2.920(98)		39	Ti	$(\frac{1}{2} \frac{3}{2} \frac{3}{2})$	0.571(64)	1.500(64)	1.349(111)
	8	O	(0 1 3)	0.062(20)	0.888(20)	3.015(93)		40	O	$(\frac{1}{2} 1 \frac{3}{2})$	0.472(59)	1.013(59)	1.487(204)
	9	O	(1 1 3)	1.378(67)	1.037(67)	2.798(20)		41	O	$(0 \frac{3}{2} \frac{3}{2})$	−0.038(20)	1.500(20)	1.552(20)
	10	Ti	$(1 \frac{3}{2} 3)$	0.988(58)	1.500(58)	3.005(62)		42	Ti	$(\frac{3}{2} \frac{3}{2} \frac{3}{2})$	1.614(59)	1.500(59)	1.440(83)
	11	O	$(\frac{1}{2} \frac{3}{2} 3)$	0.527(20)	1.500(20)	2.746(20)		43	O	$(\frac{3}{2} 1 \frac{3}{2})$	1.415(67)	1.005(67)	1.482(20)
	12	O	$(\frac{3}{2} \frac{3}{2} 3)$	1.431(34)	1.500(34)	3.226(20)		44	O	$(1 \frac{3}{2} \frac{3}{2})$	1.040(20)	1.500(20)	1.598(204)
2	13	Ti	$(\frac{1}{2} \frac{1}{2} \frac{5}{2})$	0.645(14)	0.500(14)	2.238(129)	5	45	Sr	(0 0 1)	−0.032(78)	0.003(78)	1.010(76)
	14	O	$(\frac{1}{2} 0 \frac{5}{2})$	0.413(30)	−0.071(30)	2.436(20)		46	O	$(\frac{1}{2} \frac{1}{2} 1)$	0.454(59)	0.500(59)	0.986(88)
	15	O	$(0 \frac{1}{2} \frac{5}{2})$	−0.045(38)	0.500(38)	2.384(108)		47	Sr	(1 0 1)	0.994(77)	−0.001(77)	1.012(96)
	16	Ti	$(\frac{3}{2} \frac{1}{2} \frac{5}{2})$	1.584(48)	0.500(48)	2.642(14)		48	O	$(\frac{3}{2} \frac{1}{2} 1)$	1.488(97)	0.500(97)	1.014(60)
	17	O	$(\frac{3}{2} 0 \frac{5}{2})$	1.473(61)	0.016(61)	2.499(70)		49	Sr	(0 1 1)	−0.032(78)	0.997(78)	1.010(76)
	18	O	$(1 \frac{1}{2} \frac{5}{2})$	0.925(20)	0.500(20)	2.437(204)		50	O	$(\frac{1}{2} \frac{3}{2} 1)$	0.422(20)	1.500(20)	1.020(100)
	19	Ti	$(\frac{1}{2} \frac{3}{2} \frac{5}{2})$	0.427(44)	1.500(44)	2.343(14)		51	Sr	(1 1 1)	0.994(77)	1.001(77)	1.012(96)
	20	O	$(\frac{1}{2} 1 \frac{5}{2})$	0.413(30)	1.071(30)	2.436(20)		52	O	$(\frac{3}{2} \frac{3}{2} 1)$	1.539(45)	1.500(45)	1.064(20)
	21	O	$(0 \frac{3}{2} \frac{5}{2})$	−0.025(20)	1.500(20)	2.607(90)	6	53	Ti	$(\frac{1}{2} \frac{1}{2} \frac{1}{2})$	0.504(41)	0.500(41)	0.541(41)
	22	Ti	$(\frac{3}{2} \frac{3}{2} \frac{5}{2})$	1.479(100)	1.500(100)	2.631(158)		54	O	$(\frac{1}{2} 0 \frac{1}{2})$	0.444(31)	−0.004(31)	0.538(31)
	23	O	$(\frac{3}{2} 1 \frac{5}{2})$	1.473(61)	0.984(61)	2.499(70)		55	O	$(0 \frac{1}{2} \frac{1}{2})$	0.012(31)	0.500(31)	0.528(31)
	24	O	$(1 \frac{3}{2} \frac{5}{2})$	0.953(79)	1.500(79)	2.518(20)		56	Ti	$(\frac{3}{2} \frac{1}{2} \frac{1}{2})$	1.516(41)	0.500(41)	0.535(41)
3	25	Sr	(0 0 2)	−0.012(81)	−0.006(81)	2.018(149)		57	O	$(\frac{3}{2} 0 \frac{1}{2})$	1.475(31)	−0.004(31)	0.528(31)
	26	O	$(\frac{1}{2} \frac{1}{2} 2)$	0.424(37)	0.500(37)	1.862(20)		58	O	$(1 \frac{1}{2} \frac{1}{2})$	0.994(31)	0.500(31)	0.532(31)
	27	Sr	(1 0 2)	0.935(98)	0.006(98)	2.039(100)		59	Ti	$(\frac{1}{2} \frac{3}{2} \frac{1}{2})$	0.516(41)	1.500(41)	0.514(41)
	28	O	$(\frac{3}{2} \frac{1}{2} 2)$	1.481(79)	0.500(79)	1.964(20)		60	O	$(\frac{1}{2} 1 \frac{1}{2})$	0.444(31)	1.004(31)	0.538(31)
	29	Sr	(0 1 2)	−0.012(81)	1.006(81)	2.018(149)		61	O	$(0 \frac{3}{2} \frac{1}{2})$	0.004(31)	1.500(31)	0.555(31)
	30	O	$(\frac{1}{2} \frac{3}{2} 2)$	0.356(46)	1.500(46)	1.973(20)		62	Ti	$(\frac{3}{2} \frac{3}{2} \frac{1}{2})$	1.530(41)	1.500(41)	0.509(41)
	31	Sr	(1 1 2)	0.935(98)	0.994(98)	2.039(100)		63	O	$(\frac{3}{2} 1 \frac{1}{2})$	1.475(31)	1.004(31)	0.528(31)
	32	O	$(\frac{3}{2} \frac{3}{2} 2)$	1.503(204)	1.500(204)	2.064(22)		64	O	$(1 \frac{3}{2} \frac{1}{2})$	0.993(31)	1.500(31)	0.566(31)

as consisting of alternately flipped DL (2×1) surface cells, and hence their similar chemistry explains their comparable surface energies.

The other low-energy configuration is a bulk (1×1), whereas the experimentally found DL (1×1) structure is significantly higher in energy. As we will show later, this may be explained by a (possibly dynamic) order-disorder transition. Our attempts to model the terminations against a wet medium instead of vacuum led to energies that were about 15% lower (see column "e-wet" in Table VI).

The surface energies of the two most promising models for the (2×2) DL structures were calculated [see Figs. 5(a) and 5(b)]. Although the top AL of both models differ only by a lateral shift of (1/2 1/2) bulk STO cell, the surface energy of model (a) is lower (see LDA-PWC column, Table VI). This is because a (1/2 1/2) shift means that the bonding to the next AL below is fundamentally different.

Heifets et al. suggested recently that the surface of STO in the thermodynamical equilibrium should be terminated with SrO.[59] Experimentally, such SrO terminations are typically obtained if the sample is not chemically treated (in a buffered HF solution), but annealed at temperatures above 1000 °C for 24 h to several days in oxygen.[31,60,61] In our DFT work, we did not calculate surface energies for SrO terminations. Comparing the TiO_2-terminated surfaces in Table VI, we see that the energies are in very good agreement with our values calculated using the LDA-PWC functional. Thus, if thermodynamical equilibrium is not reached, the presence of kineti-

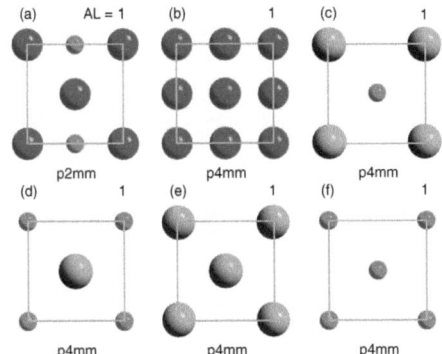

FIG. 8. (Color) Hot (1×1) starting models using the same representation as in Fig. 3.

cally favored TiO$_2$-terminated surfaces remains an interesting suggestion and is compatible with the experimental observations.

Our DFT results indicate that the (2×1) DL reconstruction should be equally favored as bulk-terminated STO (in contrast to the results presented by Johnston et al.,[17]), and that the (2×2) DL reconstruction should be most favored.

If we compare the energies obtained using the GGE-PBE functional with the results of Warschkow et al.,[52] we see a consistent trend for the difference between the (1×1)-DL surface energy and those of the (2×1)-DL and (2×2)-DL reconstructions, which confirms the very low surface energies for the (2×1) and (2×2) DL reconstructions. It is

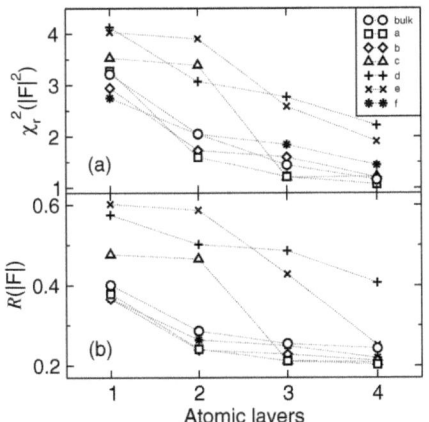

FIG. 9. The evolution of χ_r^2 (a) and R factor (b) as a function of the fit ALs for hot STO under identical conditions and started from the same initial positions. The dotted lines are guides to the eyes.

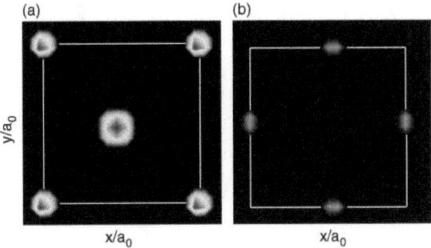

FIG. 10. (Color online) Cuts through the average electron density recovered by PARADIGM for coherently added domains at two different heights: O positions at $z/a_0=2.97$ (a) and Ti positions at $z/a_0=3.28$ (b). Representation is in bulk units of STO and analogously to Table V, for direct comparison. The white line indicates the STO bulk unit cell.

noted that the (2×2) reconstruction observed in this work is among the three surfaces with the lowest absolute energy. The $c(4×2)$ has been experimentally observed by annealing in (i) O$_2$ at temperatures of 850–930 °C,[62] (ii) H$_2$ at 950 °C,[10] and (iii) in UHV after Ar ion sputtering at a temperature of 1200 °C,[15] while the $(\sqrt{2}×\sqrt{2})R45°$ reconstruction has yet to be experimentally observed.

D. Formation of reconstructions

A possible explanation for the driving force of the formation of the reconstructions could invoke the role of Ca im-

TABLE V. Refined positions and DW factors (expressed as root mean squared displacements in brackets) for the (1×1) reconstruction under hot conditions in units of bulk STO.

AL	No.	Atom	Nominal position	x/a_0	y/a_0	z/a_0
1	1	Ti	$(\tfrac{1}{2}\ 0\ 3)$	0.500(204)	0.000(204)	3.256(204)
	2	O	$(0\ 0\ 3)$	0.000(71)	0.000(71)	2.990(29)
	3	O	$(\tfrac{1}{2}\ \tfrac{1}{2}\ 3)$	0.500(204)	0.500(204)	3.102(204)
2	4	Ti	$(\tfrac{1}{2}\ \tfrac{1}{2}\ \tfrac{5}{2})$	0.500(35)	0.500(35)	2.574(118)
	5	O	$(\tfrac{1}{2}\ 0\ \tfrac{5}{2})$	0.500(125)	0.000(125)	2.616(22)
	6	O	$(0\ \tfrac{1}{2}\ \tfrac{5}{2})$	0.000(20)	0.500(20)	2.485(62)
3	7	Sr	$(0\ 0\ 2)$	0.000(45)	0.000(45)	2.009(44)
	8	O	$(\tfrac{1}{2}\ \tfrac{1}{2}\ 2)$	0.500(143)	0.500(143)	1.929(41)
4	9	Ti	$(\tfrac{1}{2}\ \tfrac{1}{2}\ \tfrac{3}{2})$	0.500(27)	0.500(27)	1.506(14)
	10	O	$(\tfrac{1}{2}\ 0\ \tfrac{3}{2})$	0.500(22)	0.000(22)	1.545(20)
	11	O	$(0\ \tfrac{1}{2}\ \tfrac{3}{2})$	0.000(20)	0.500(20)	1.410(20)
5	12	Sr	$(0\ 0\ 1)$	0.000(36)	0.000(36)	1.004(41)
	13	O	$(\tfrac{1}{2}\ \tfrac{1}{2}\ 1)$	0.500(51)	0.500(51)	0.947(44)
6	14	Ti	$(\tfrac{1}{2}\ \tfrac{1}{2}\ \tfrac{1}{2})$	0.500(26)	0.500(26)	0.502(26)
	15	O	$(\tfrac{1}{2}\ 0\ \tfrac{1}{2})$	0.500(43)	0.000(43)	0.510(43)
	16	O	$(0\ \tfrac{1}{2}\ \tfrac{1}{2})$	0.000(43)	0.500(43)	0.474(43)

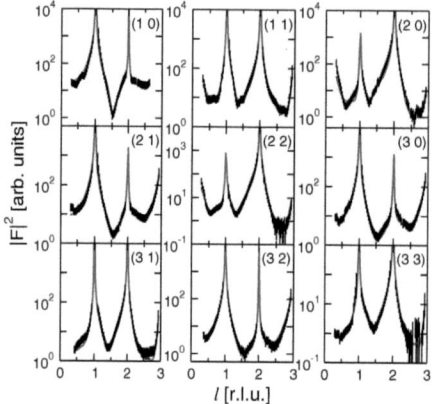

FIG. 11. (Color online) Set of the SXRD data (black data points) and calculated intensities using the TiO$_2$-DL model (gray solid line, red online) for the hot structure.

purities. On the one hand, these have been demonstrated to be of great importance to reconstruct and modify surfaces for a variety of oxides, including SrO-terminated STO(001),[6] MgO(001),[63,64] TiO$_2$(110),[65] and Fe$_3$O$_4$(001).[66] On the other hand, trace amounts of Ca are a source of impurities in the production of single-crystalline STO.[22]

Of particular interest in this context is the work by Andersen and Møller,[6] where they used Auger electron spectroscopy and low-energy electron-diffraction to observe the formation of an overlayer structure (with Ca incorporated in the surface) for a Ca content of 3%, and a (2×2) reconstruction when the Ca content was below 1%. This reconstruction later disappeared completely when the Ca was fully removed. It is noteworthy that they judged this (2×2) reconstruction to penetrate at most 3 to 4 MLs into the bulk.

The LEIS spectra shown in Fig. 2 suggest a similar behavior of TiO$_2$-terminated STO: As-received samples contain a higher amount of Ca at the surface and show no reconstructions. The chemical and thermal treatment of the surface according to Refs. 23 and 24 may lead to a decreased Ca content in the surface that allows the formation of reconstructions, as observed by us. A further etch cycle, as suggested by Ohnishi et al.,[27] diminishes the Ca content further and may result in the complete absence of reconstructions again. This is, indeed, what we observed: Not all of the re-etched STO samples showed reconstructions. There seems to be a direct correlation between the presence of reconstructions and the Ca impurities, the amount of the latter determined by both the intrinsic Ca content of STO single crystals and the etching conditions.

The LEIS spectra show further an amount of Sr present in the outermost layer (see Fig. 2). Here one has to take into account that the Sr signal compared to Ti is enhanced by a factor of 1.7 due to the different mass of the nuclei that change their detection probabilities accordingly.[67] Moreover, even a modest treatment of the surface with pure water or buffered HF removes residing Sr contaminations.[27] We note that in our case the re-etch procedure did not remove all the residuals on the sample on which LEIS was performed.[68] The green spectrum in Fig. 2 shows among Sr also a significant F peak, an impurity attesting to possibly incomplete purification. The F peak is likely due to the buffer salt (NH$_4$F). Although the samples we used for SXRD did not undergo an additional etching step to remove any residuals, it seems likely that the use of highly brilliant synchrotron radiation has the same effect. In any case, our attempts to model the surface with additional Sr at the surface failed.

TABLE VI. DFT surface energies in eV/(1×1) unit cell (with respect to the SrO chemical potential) for selected models. The letters in brackets identify the models in the corresponding figures.

	Present work				Heifets[a]	Johnston[b]	Warschkow[c]
Functional STO bulk unit cells	LDA-PWC 6	GGA-PBE 6	e-wet 6	GGA-PBE 4	B3PW-PP 4	LDA-?[d] 4	GGA-91 2
(1×1) bulk, (a) and (b)	1.02	0.78	0.67		1.05	1.28	
(1×1) DL (c)	1.64	1.39	1.26		1.75		0[e]
(2×1) DL (a)	1.31	0.78	0.63	0.81	1.3	2.00	−0.10
(2×2) DL (a)	0.79	0.38	0.33				−0.66
(2×2) DL (b)	1.22						−0.16
c(4×2)				0.49			−0.64
($\sqrt{2}\times\sqrt{2}$)R45°				0.33			−0.79

[a]Reference 60 from Fig. 6(a) therein.
[b]Reference 17 from Figs. 4 and 5(a) therein.
[c]Reference 52.
[d]Exact functional not specified.
[e]Subsequent values refer to this configuration as having zero energy.

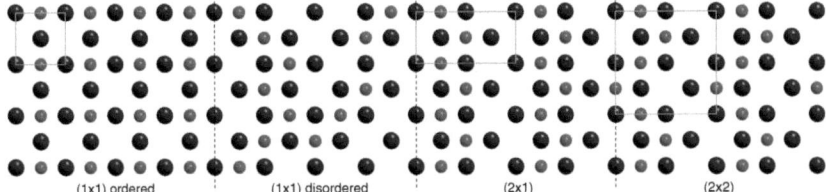

FIG. 12. (Color online) A schematic showing the similarity and simple transformation between the top layers of either the (ordered) (1×1), (2×1), and (2×2) terminations. Transformation of the ordered (1×1)-DL structure and either reconstruction only requires a diagonal movement of every second Ti atom, with the possibility of forming a disordered (1×1) relaxation ("grain boundaries") in between the domains. The gray boxes indicate the surface unit cells of the terminations. Color code: Ti small red, O large blue circles.

E. Order-disorder transition

Transformation between the DL (1×1) relaxation and either reconstruction only requires a diagonal hop of every second Ti atom across half a surface (1×1) unit cell, as schematically shown in Fig. 12. The surface energy of the (1×1) DL structure seems, however, anomalously high (see Table VI). The experimentally determined presence of a (1×1) structure could have two reasons: either this (1×1) domain is metastable with a significant activation barrier to lower-energy states, or in fact it consists of a disordered mixture of the characteristic zigzag motif of the (2×1) and (2×2), randomly flipped and mirrored.[69] This (possibly dynamically) disordered (1×1) domain would have a surface energy approximately as low as the two reconstructions, and the average structure observed via SXRD could be described by the TiO$_2$ DL (1×1). Such a model would not have to invoke coordinated and concerted Ti-hopping across the surface over large distances (of the order of some 100 nm) to go from the reconstructions to the (1×1) relaxation. Disorder could also explain how such a (1×1) surface can dominate at elevated temperatures, where the surface thermal vibrational energy becomes comparable to the activation barrier between the (2×1) and (2×2), leading to a disordered mixture. Indeed, the difference in surface energy between two (2×1) cells and a single (2×2) cell is $\Delta E_s = 1.6$ eV (see Table VI). On the other hand, the vibrational energy of the surface atoms of the same system is, to a first approximation, equal to $E_v = 3kT \times (4n_a)$, where k is Boltzmann's constant, T is the temperature, n_a is the number of atoms per top sublayer and unit cell (here, for TiO$_2$, $n_a=3$), and the factor 4 accounts for the fact that 4 surface (1×1) cells are needed to describe this system. We therefore obtain $E_v = 3.2$ eV under hot conditions, i.e., twice the difference in reconstruction surface energies. The size of the activation barrier between the two reconstructions will affect the time needed to reach this mixed equilibrium state. From the Arrhenius rate constant of the disappearance of the SSR signal of the order of $k_s = 0.01$ s^{-1}, and assuming a preexponential factor $A \sim 10^{12}$, we obtain an activation energy of $E_a \approx 3$ eV. For the room temperature sample, $E_v = 0.9$ eV and the reaction rate constant is of the order of 10^{-40} s^{-1}, i.e., the system is completely kinetically hindered. On a speculative note, the disordered (1×1) domains suggested here may, in the cold sample, comprise regions, or "grain boundaries" of disorder between the ordered (2×1) and (2×2) reconstructions, frozen in as the sample cools at the end of the substrate preparation procedure.

F. Electrostatic considerations

1. Atomic displacements

In Fig. 13, we show the displacements of the atomic positions Δz in the final model compared to the high-symmetry positions in the starting models. The displacements are most prominent in the z direction (i.e., normal to the surface), and, for the (1×1) relaxations, movements only along this axis are allowed, due to the surface symmetry.

Our data show significant atomic movements for all refined layers down to a depth of three unit cells, the uppermost atoms on average displaced outwards to the vacuum. This implies a loss of centrosymmetry of the surface Ti-atoms within the O octahedra and may lead to a permanent

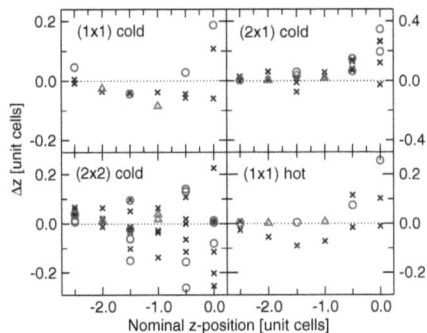

FIG. 13. (Color online) Displacements of Sr (green triangles), Ti (red circles), and O (blue crosses) in the z direction from the high-symmetry positions in the starting models. The nominal surface is at $z=0$ and positive values of Δz indicate displacements towards vacuum.

surface dipole moment and surface ferroelectricity.[2,70,71]

The trends shown in our experimental data for the (2×1) DL structure are very similar to those theoretically predicted by Johnston et al.[17] Therefore, theoretical and experimental arguments both support the puckering of the uppermost Ti sites in STO.

We note that the atomic movements in the z direction for the (2×1) and the (2×2) reconstruction are different. This is a surprising result under consideration of the chemical similarity and the proposed formation mechanism of the two reconstructions. With respect to Fig. 7 and Tables III and IV, we see more pronounced movements of the atoms in the xy plane of the (2×2) reconstruction. A possible explanation could involve symmetry arguments as follows: The pm symmetry applied to the (2×2) unit cell allows movements in the y direction for certain atoms whereas this symmetry assigned to a (2×1) reconstruction prohibits such shifts completely. Therefore, the atoms of the (2×2) reconstruction are less constrained to movements out of the surface plane, but the additional freedom of y movements may help obtaining a low surface energy.[72]

2. Polar surfaces

According to classical electrostatics, the stability of a crystal surface depends on the charge distribution in the unit cell in the direction perpendicular to the surface. Polar oxide surfaces can be classified into three different types.[73] Type 1 surfaces have neither a net charge σ nor a net dipole moment μ perpendicular to the surface. A typical example would be the MgO(001) surface that consists of layers containing equal amounts of Mg and O atoms. The type 2 surface displays a nonzero charge $\sigma \neq 0$, but has no net dipole moment within the unit cell in the normal direction, $\mu = 0$, e.g., a $TiO_2(110)$ surface. Because $\mu=0$, these two surface types are called nonpolar and are potentially stable. In a polar type 3 surface, $\sigma \neq 0$ and $\mu \neq 0$ in the repeat units of the structure. An example of a type 3 surface is MgO(111). Assuming ionic bonding, the total dipole moment of repeated units normal to the surface is directly proportional to the slab thickness, and therefore, the surface energy per unit area diverges even for thin films. This is the origin of the surface instability of type 3 surfaces.

An STO(001) perovskite crystal consists of repeated TiO_2 and SrO layers, and one might assume it to be type 1, because the formal charges of Ti^{4+}, Sr^{2+}, and O^{2-} are compensated layer by layer and no net dipole moment is present. However, STO is not fully ionic. The partially covalent nature of the bonds (a property later referred to as "back transfer") in STO results in ALs having nonzero and opposite net charges $\pm\sigma$.[51] Thus, STO(001) is a "weakly polar" type 3 surface. For polar systems with only a single distorted layer, this top AL is required to have a charge of $\sigma' = \sigma/2$ in order to avoid instability.

One can now calculate the net charge per unit area for a bulk-coordinated unit cell of STO using a covalency-induced back transfer of 0.422 electrons (e^-) per Ti and $0.045e^-$ per Sr site, respectively.[74] Taking the coordination of the different atoms into account, this results in a net charge per unit cell area of $\sigma a_0^2 = -0.484e^-$ for a TiO_2 AL that is compensated by $\sigma a_0^2 = +0.484e^-$ of the SrO AL, i.e., STO is indeed a polar system. The terminating TiO_2 layer of bulk STO has to meet the condition of $\sigma' = \sigma/2$ to be stable. Simple calculations show that for the reduced number of bonds of a topmost TiO_2 layer $\sigma' a_0^2 = -0.242e^-$.

The addition of a second TiO_2 layer in the DL model modifies the charge of the original TiO_2 AL. We can also calculate the DL structures in order to show that this TiO_2 overlayer compensates exactly in all three DL surface structures for this change. For all three surface terminations, we can neglect the bulk net charges as we showed the TiO_2 and SrO layers to compensate for each other before, and therefore, the top three ALs (i.e., one SrO and two TiO_2 layers) solely remain of interest. Let us start from the SrO layer and move towards the surface. For all three terminations, the SrO layer is bulk-coordinated which leads to a net charge of $\sigma a_0^2 = +0.484e^-$ per STO unit cell area. The following TiO_2 layer is coordinated differently compared to a bulk TiO_2 layer, and the coordination depends on the structure of the terminating surface layer. However, the resulting net charge for the TiO_2 layer under observation is the same for (1×1), (2×1), and (2×2) domains, and thus $\sigma' a_0^2 = -0.242e^-$, already meeting the electrostatic stability criterion. Therefore, the top TiO_2 layer should have no net charge in order to be stable. Calculating this net charge for every TiO_2 layer of the (1×1), (2×1), and (2×2) individually, we indeed obtain $\sigma'' a_0^2 = 0$ for every termination. As a consequence, the (1×1), (2×1), and (2×2) termination all satisfy the electrostatic requirements. Note, however, that this criterion is met neither by the Sr adatom model nor the O overlayer model. If the metal-rich surfaces suggested by our direct methods result in a metallic layer, the required compensation could occur, but the electronic structure of such an overlayer is not known.

IV. CONCLUSIONS

We have presented a detailed and broad study of the surface structure of the technologically and scientifically important perovskite material $SrTiO_3(001)$ using SXRD, both at room temperature in UHV and at conditions typical for the growth of thin films. The surface was also characterized $ex\ situ$ by AFM, XPS, and LEIS, and a TiO_2 termination was found.

The cold surface simultaneously contains three different terminations, namely, (2×1) and (2×2) reconstructions and a (1×1) relaxation. During the refinement of the cold data set, over 70 models were tested, of which 49 are described in more detail in this work. The atomic coordinates of the final model are given. It consists of a characteristic double TiO_2 top layer, whereby the two reconstructions contain a repetition of a distinctive zigzag motif, similar to the one proposed by Erdman et al. for the (2×1) termination.[14] Both reconstructions are energetically favorable according to our DFT calculations, in agreement with other theoretical work.[52,59]

The hot surface can be modelled using a double-layered

TiO$_2$ termination, very similar to the cold (1×1), but with more pronounced puckering, and the atomic coordinates are shown. Direct methods analysis using PARADIGM supports the TiO$_2$-rich surface and the atomic positions.

Both the cold and hot surfaces have significant deviations from the high-symmetry starting positions down to a depth of three unit cells. This may have important consequences regarding surface ferroelectricity and other nonlinear properties of the surface.

Surface vibrational energy considerations suggest the possibility of the (1×1) in fact being a temperature-induced disordered mixture of the two types of reconstructions. This would explain the presence of only the (1×1) structure upon heating the sample. Finally, our experimental results are in agreement with electrostatic considerations on the stability of the polar surface of STO.

ACKNOWLEDGMENTS

Fruitful discussions with R. Feidenhans'l, I. K. Robinson, H. H. Brongersmaa, M. de Ridder, R. ter Veen, K. Peters, J. Krempasky, and L. Patthey are gratefully acknowledged. Support of this work by the Schweizerischer Nationalfonds zur Förderung der wissenschaftlichen Forschung and the staff of the Swiss Light Source is gratefully acknowledged. This work was partly performed at the Swiss Light Source, Paul Scherrer Institut. Work at UWM is supported by the US DOE under contract No. DE-FG02-06ER46277.

---

*philip.willmott@psi.ch

[1] S. Jin, T. H. Tiefel, M. McCormack, R. A. Fastnacht, R. Ramesh, and L. H. Chen, Science **264**, 413 (1994).
[2] D. D. Fong, G. B. Stephenson, S. K. Streiffer, J. A. Eastman, O. Auciello, P. H. Fuoss, and C. Thompson, Science **304**, 1650 (2004).
[3] C. H. Ahn, K. M. Rabe, and J.-M. Triscone, Science **303**, 488 (2004).
[4] A. Ohtomo and H. Y. Hwang, Nature (London) **427**, 423 (2004).
[5] J. H. Haeni et al., Nature (London) **430**, 758 (2004).
[6] J. E. T. Andersen and P. J. Møller, Appl. Phys. Lett. **56**, 1847 (1990).
[7] T. Matsumoto, H. Tanaka, T. Kawai, and S. Kawai, Surf. Sci. **278**, L153 (1992).
[8] M. Naito and H. Sato, Physica C **229**, 1 (1994).
[9] Q. D. Jiang and J. Zegenhagen, Surf. Sci. **338**, L882 (1995).
[10] Q. D. Jiang and J. Zegenhagen, Surf. Sci. **425**, 343 (1999).
[11] G. Charlton, S. Brennan, C. A. Muryn, R. McGrath, D. Norman, T. S. Turner, and G. Thornton, Surf. Sci. **457**, L376 (2000).
[12] T. Kubo and H. Nozoye, Phys. Rev. Lett. **86**, 1801 (2001).
[13] A. Kazimirov, D. M. Goodner, M. J. Bedzyk, J. Bai, and C. R. Hubbard, Surf. Sci. **492**, L711 (2001).
[14] N. Erdman, K. R. Poeppelmeier, O. Warschkow, D. E. Ellis, and L. D. Marks, Nature (London) **419**, 55 (2002).
[15] M. R. Castell, Surf. Sci. **505**, 1 (2002).
[16] M. R. Castell, Surf. Sci. **516**, 33 (2002).
[17] K. Johnston, M. R. Castell, A. T. Paxton, and M. W. Finnis, Phys. Rev. B **70**, 085415 (2004).
[18] V. Vonk, S. Konings, G. J. van Hummel, S. Harkema, and H. Graafsma, Surf. Sci. **595**, 183 (2005).
[19] F. Silly, D. T. Newell, and M. R. Castell, Surf. Sci. **600**, L219 (2006).
[20] L. M. Liborio, C. G. Sánchez, A. T. Paxton, and M. W. Finnis, J. Phys.: Condens. Matter **17**, L223 (2005).
[21] R. Herger, P. R. Willmott, O. Bunk, C. M. Schlepütz, B. D. Patterson, and B. Delley, Phys. Rev. Lett. **98**, 076102 (2007).
[22] R. Herger and K. Peters (private communication).
[23] M. Kawasaki, K. Takahashi, T. Maeda, R. Tsuchiya, M. Shinohara, O. Ishiyama, T. Yonezawa, M. Yoshimoto, and H. Koinuma, Science **266**, 1540 (1994).
[24] G. Koster, B. L. Kropman, G. J. H. M. Rijnders, D. H. A. Blank, and H. Rogalla, Appl. Phys. Lett. **73**, 2920 (1998).
[25] J. C. Dupin, D. Gonbeau, P. Vinatier, and A. Levasseur, Phys. Chem. Chem. Phys. **2**, 1319 (2000).
[26] H. H. Brongersma, M. Draxler, M. de Ridder, and P. Bauer, Surf. Sci. Rep. **62**, 63 (2007).
[27] T. Ohnishi, K. Shibuya, M. Lippmaa, D. Kobayashi, H. Kumigashira, M. Oshima, and H. Koinuma, Appl. Phys. Lett. **85**, 272 (2004).
[28] B. D. Patterson et al., Nucl. Instrum. Methods Phys. Res. A **540**, 42 (2005).
[29] P. R. Willmott et al., Appl. Surf. Sci. **247**, 188 (2005).
[30] C. M. Schlepütz, R. Herger, P. R. Willmott, B. D. Patterson, O. Bunk, C. Brönnimann, B. Henrich, G. Hülsen, and E. F. Eikenberry, Acta Crystallogr., Sect. A: Found. Crystallogr. **61**, 418 (2005).
[31] K. Szot and W. Speier, Phys. Rev. B **60**, 5909 (1999).
[32] E. Vlieg, J. Appl. Crystallogr. **30**, 532 (1997).
[33] O. Bunk, Ph.D. thesis, University of Hamburg, Department of Physics, 1999, URL http://www.sub.uni-hamburg.de/opus/volltexte/1999/99/
[34] W. C. Hamilton, Acta Crystallogr. **18**, 502 (1965).
[35] The R factor used here is defined as $R(|F|) = \frac{\sum_{i=1}^{N}||F|_{i,\exp}-|F|_{i,\text{theor}}|}{\sum_{i=1}^{N}|F|_{i,\exp}}$, where $|F|$ is the amplitude of the structure factor.
[36] Reduced χ_r^2, values were used as defined by $\chi_r^2(|F|^2) = \frac{1}{N-P}\sum_{i=1}^{N}\frac{(|F|^2_{i,\exp}-|F|^2_{i,\text{theor}})^2}{\sigma^2_{|F|^2_{\exp}}}$, where N is the number of measured data points, P the number of parameters, $|F|^2$ the squared amplitude of the structure factor, i.e., the measured intensity, and $\sigma_{|F|^2}$, the error of the measured intensity.
[37] K. Lonsdale, in *International Tables for X-Ray Crystallography*, edited by C. H. MacGillavry, G. D. Rieck, and K. Lonsdale (Reidel, Dordrecht, 1985), Vol. III, Chap. 3.3.5, p. 232.
[38] K. N. Trueblood, H. B. Burgi, H. Burzlaff, J. D. Dunitz, C. M. Gramaccioli, H. H. Schulz, U. Shmueli, and S. C. Abrahams, Acta Crystallogr., Sect. A: Found. Crystallogr. **A52**, 770 (1996).
[39] B. Delley, J. Chem. Phys. **92**, 508 (1990).
[40] B. Delley, J. Phys. Chem. **100**, 6107 (1996).
[41] B. Delley, J. Chem. Phys. **113**, 7756 (2000).
[42] J. P. Perdew and Y. Wang, Phys. Rev. B **45**, 13244 (1992).
[43] J. P. Perdew, K. Burke, and M. Ernzerhof, Phys. Rev. Lett. **77**, 3865 (1996).

[44] B. Delley, Mol. Simul. **32**, 117 (2006).
[45] E. Vlieg, J. Appl. Crystallogr. **33**, 401 (2000).
[46] X. Torrelles, J. Rius, F. Boscherini, S. Heun, B. H. Mueller, S. Ferrer, J. Alvarez, and C. Miravitlles, Phys. Rev. B **57**, R4281 (1998).
[47] Y. Yacoby, R. Pindak, R. MacHarrie, L. Pfeiffer, L. Berman, and R. Clarke, J. Phys.: Condens. Matter **12**, 3929 (2000).
[48] D. K. Saldin, R. J. Harder, V. L. Shneerson, and W. Moritz, J. Phys.: Condens. Matter **13**, 10689 (2001).
[49] L. D. Marks, N. Erdman, and A. Subramanian, J. Phys.: Condens. Matter **13**, 10677 (2001).
[50] J. R. Fienup, Appl. Opt. **21**, 2758 (1982).
[51] C. Noguera, J. Phys.: Condens. Matter **12**, R367 (2000).
[52] O. Warschkow, M. Asta, N. Erdman, K. R. Poeppelmeier, D. E. Ellis, and L. D. Marks, Surf. Sci. **573**, 446 (2004).
[53] P. N. Keating, Phys. Rev. **145**, 637 (1966).
[54] See EPAPS Document No. E-PRBMDO-76-155739 for 3D animations of the final structures and tabulated atomic coordinates in ASCII format. For more information on EPAPS, see http://www.aip.org/pubservs/epaps.html
[55] Note that the FIT refinement assumed incoherently added domains. For this particular model of the hot data set, however, we also tested coherent domain addition and got almost exactly the same result as for incoherent addition.
[56] D. K. Saldin, R. J. Harder, V. L. Shneerson, and W. Moritz, J. Phys.: Condens. Matter **14**, 4087 (2002).
[57] R. Fung, V. L. Shneerson, P. F. Lyman, S. S. Parihar, H. T. Johnson-Steigelman, and D. K. Saldin, J. Phys.: Condens. Matter **63**, 239 (2007).
[58] D. K. Saldin and V. L. Shneerson, J. Phys.: Condens. Matter (to be published).
[59] E. Heifets, S. Piskunov, E. A. Kotomin, Y. F. Zhukovskii, and D. E. Ellis, Phys. Rev. B **75**, 115417 (2007).

[60] Y. Liang and D. A. Bonnell, Surf. Sci. **310**, 128 (1994).
[61] Szot and Speier attribute the presence of droplet-like features in their AFM pictures to the formation of SrO islands (Ref. 31). Note that our AFM image in Fig. 1 shows no such droplet features.
[62] N. Erdman, O. Warschkow, M. Asta, K. R. Poeppelmeier, D. E. Ellis, and L. D. Marks, J. Am. Chem. Soc. **125**, 10050 (2003).
[63] M. Gajdardziska-Josifovska, P. A. Crozier, and M. R. McCartney, Surf. Sci. **284**, 186 (1993).
[64] K. Fukui and Y. Iwasawa, Surf. Sci. **441**, 529 (1999).
[65] L. P. Zhang and U. Diebold, Surf. Sci. **413**, 242 (1998).
[66] G. Mariotto, S. Murphy, N. Berdunov, S. F. Ceballos, and I. V. Shvets, Surf. Sci. **564**, 79 (2004).
[67] B. D. Patterson, R. Herger, H. H. Brongersma, M. de Ridder, and R. ter Veen (private communication).
[68] Discussions with Calipso B. V. made clear that it is difficult to identify surface impurities and Sr residuals in the outermost layer by coaxial impact-collision iron scattering spectroscopy, the technique used in the work by Ohnishi and co-workers (Ref. 27 and 67).
[69] L. D. Roelofs, G. Y. Hu, and S. C. Ying, Phys. Rev. B **28**, 6369 (1983).
[70] N. Bickel, G. Schmidt, K. Heinz, and K. Müller, Phys. Rev. Lett. **62**, 2009 (1989).
[71] V. Ravikumar, D. Wolf, and V. P. Dravid, Phys. Rev. Lett. **74**, 960 (1995).
[72] On a speculative note, the pm, symmetry allowing in-plane movements along the y direction might be a reason for the significantly lower surface energy of the (2×2), compared to the (2×1), reconstruction in our DFT calculations (see Table VI), although the chemical similarity of the reconstructions would not necessarily point to such a distinct energy difference.
[73] P. Tasker, J. Phys. C **12**, 4977 (1979).
[74] J. Goniakowski and C. Noguera, Surf. Sci. **365**, L657 (1996).

Paper III

Pulsed laser deposition of atomically flat $La_{1-x}Sr_xMnO_3$ thin films using a novel target geometry

The work presented in this chapter has been published in:
P.R. Willmott, R. Herger, M.C. Falub, L. Patthey, M. Döbeli, C.V. Falub, M. Shi, and M. Schneider, *Pulsed laser deposition of atomically flat $La_{1-x}Sr_xMnO_3$ thin films using a novel target geometry*, Appl. Phys. A **79**, 1199 (2004).

artikel3.pdf is the online version of the library (Zentralbibliothek Zürich).

Abstract

A new ablation target geometry is presented that was used to produce thin films of $La_{1-x}Sr_xMnO_3$ grown heteroepitaxially on $SrTiO_3$ by pulsed reactive crossed-beam laser ablation. The films were grown in order to perform angle-resolved photoelectron spectroscopy, which demands that the surface be atomically flat. In situ and ex situ analysis shows that this condition was met, even after depositing to a thickness of over 100 nm.

DOI: 10.1007/s00339-004-2717-8 [a]
PACS numbers: 61.10.-i, 61.18.Bn, 68.47.Gh, 71.27.+a, 81.15.Fg

Reprinted with kind permission from Springer Science and Business Media.
[a] Note that you need a subscription for this journal to directly access the article.

P.R. WILLMOTT[1,2,✉]
R. HERGER[1,2]
M.C. FALUB[1]
L. PATTHEY[1]
M. DÖBELI[3]
C.V. FALUB[4]
M. SHI[1]
M. SCHNEIDER[5]

Pulsed laser deposition of atomically flat La$_{1-x}$Sr$_x$MnO$_3$ thin films using a novel target geometry

[1] Swiss Light Source, Paul Scherrer Institute, 5232 Villigen, Switzerland
[2] Physical Chemistry Institute, University of Zürich, Winterthurerstrasse 190, 8057 Zürich, Switzerland
[3] ETH Hönggerberg, Paul Scherrer Institute, 8093 Zürich, Switzerland
[4] Laboratory for Micro- and Nanotechnology, Paul Scherrer Institute, 5232 Villigen, Switzerland
[5] Laboratory for Neutron Scattering, Paul Scherrer Institute, 5232 Villigen, Switzerland

Received: 29 September 2003/Accepted: 24 February 2004
Published online: 26 July 2004 • © Springer-Verlag 2004

ABSTRACT A new ablation target geometry is presented that was used to produce thin films of La$_{1-x}$Sr$_x$MnO$_3$ grown heteroepitaxially on SrTiO$_3$ by pulsed reactive crossed-beam laser ablation. The films were grown in order to perform angle-resolved photoelectron spectroscopy, which demands that the surface be atomically flat. In situ and ex situ analysis shows that this condition was met, even after depositing to a thickness of over 100 nm.

PACS 61.10.-i; 61.18.Bn; 68.47.Gh; 71.27.+a; 81.15.Fg

1 Introduction

Since the discovery of colossal magnetoresistance in the heavily doped perovskite manganites by Jin et al. in 1993 [1], the physics responsible for this phenomenon has been intensively investigated. A powerful analytical technique is angle-resolved photoemission spectroscopy (ARPES) [2], which can yield important information on the relevant low-binding-energy electronic states such as those responsible for the double-exchange hopping mechanism along the Mn−O chains [3]. However, the experimental determination of the electronic structure via ARPES is only possible with a perfect sample surface. This is because only the parallel component of the photoelectron's momentum vector k is preserved as it escapes the sample surface; hence the surface and crystallographic orientations relative to the detector must be well defined.

It is exceedingly difficult to prepare a clean and crystallographically unique surface by in-vacuum cleaving of perovskite structures; this is due mainly to their isotropic cubic symmetry. Preparation of high-quality heteroepitaxial films in situ is therefore an exciting alternative approach.

One of the most intriguing aspects of strongly correlated electron systems, such as La$_{1-x}$Sr$_x$MnO$_3$, is the influence of the degree of doping (x) on the electronic properties. Pulsed laser deposition (PLD) has been shown to be a very attractive film growth technique for such chemically complex materials due to its ability to faithfully transfer the stoichiometry

of the bulk target to the growing film [4]. Normally, therefore, an investigation of the effect of doping on the physical properties of the films would require a set of target materials covering the doping region of interest. In this paper, we present results for the growth of La$_{1-x}$Sr$_x$MnO$_3$ by pulsed reactive crossed-beam laser ablation (PRCLA) [5], whereby x could be chosen at will between 0 and 1 without the need to change the target [6].

2 Experimental setup

The experimental setup has been described comprehensively elsewhere [6, 7]. Briefly, La$_{1-x}$Sr$_x$MnO$_3$ thin films were deposited in an ultrahigh vacuum chamber (base pressure 7×10^{-9} Pa) by ablating a target rod composed of a section of LaMnO$_3$ (LMO) and one of SrMnO$_3$ (SMO) using the fourth harmonic of a Nd : YAG laser (10 Hz, 266 nm, 5 ns, 2 J cm^{-2}) in the presence of a synchronized N$_2$O gas pulse (average pressure 2×10^{-2} Pa). N$_2$O was chosen as the primary oxidizing gas due to its enhanced ability over O$_2$ in oxidizing Mn [8]. O$_2$ was also bled into the chamber (1.5×10^{-2} Pa) to force the equilibrium of the reaction

$$La_{1-x}Sr_xMnO_{3-y} + \frac{y}{2}O_2 \rightleftharpoons La_{1-x}Sr_xMnO_3 \quad (1)$$

toward full oxidation of the deposited film [9]. Films were deposited on 10×10 mm^2 atomically flat SrTiO$_3$(001) substrates at 750 °C.

A 100 nm film would typically require 46 000 laser shots. The doping x was set by the ablated lengths of each section l_{SMO} and l_{LMO}. The rod was translated at 28 mm s^{-1}, and $l_{SMO} + l_{LMO}$ was typically 40 mm. A complete target stroke cycle was thus completed in about 3 s (30 laser pulses), significantly less than the time required to deposit a complete monolayer. This ensured alloy growth [6].

The deposition apparatus was adapted to be compatible with the ARPES station at the Surfaces and Interfaces Spectroscopy beamline at the Swiss Light Source.

The films were analyzed in situ during and after deposition using reflection high-energy electron diffraction (RHEED). The films were analyzed ex situ for their chemical composition by Rutherford backscattering spectroscopy (RBS, 2 MeV ^{4}He$^+$ ions) and for their crystallinity using X-ray diffraction (Cu K_α radiation plus a Ge(220) double crystal monochromator). The temperature dependence of the film resistance was

✉ Fax: +41-56/310-4551, E-mail: philip.willmott@psi.ch

studied using a temperature-controlled four-point resistance setup.

3 Results and discussion

The film quality could be monitored during growth using RHEED. There was no evidence of intensity oscillations of the specular (00) reflection, though, as the two-dimensional diffraction zero-order and first-order Laue zones remained throughout the entire deposition (Fig. 1), it can be concluded that the films grew in a step flow mode. Kikuchi lines could also be seen throughout growth, which is indicative of high crystalline purity.

Heteroepitaxial growth was further confirmed ex situ by X-ray reflectivity and diffraction. The low-angle reflectivity curves showed clear Kiessig fringes that could be fitted to a film roughness of less than 1 Å, even for films grown to a thickness of 130 nm. The fringes were also seen in the $\theta - 2\theta$ spectra, which only showed (00l) reflections of the substrate and film (Fig. 2). The out-of-plane lattice constants for $La_{0.66}Sr_{0.34}MnO_3$ and $SrTiO_3$ were determined to be 3.870 Å and 3.905 Å, respectively. The film is therefore tetragonally distorted with a shortening of c compared to the bulk value of 0.15%, due to the anisotropic strain induced by the lattice mismatch in-plane of $+0.7\%$.

The RBS data are summarized in Fig. 3. In addition to allowing one to determine the film stoichiometry, channelling experiments yield information on the film quality, especially the density of crystallographic faults (e.g., edge and screw dislocations) with a sensitivity excluded to laboratory-based XRD. The films showed the maximum degree of channelling possible for the experimental setup, with a χ^2-value of only 3% (Fig. 3a). The slopes of the substrate and film peak tops are identical, which implies that their crystal quality is the same. The change in stoichiometry as a function of the ratio of ablation length of the two components of the target l_{SMO}/l_{LMO} is given by

$$x = \frac{Q_{SMO}l_{SMO}}{Q_{SMO}l_{SMO} + Q_{LMO}l_{LMO}} \qquad (2)$$

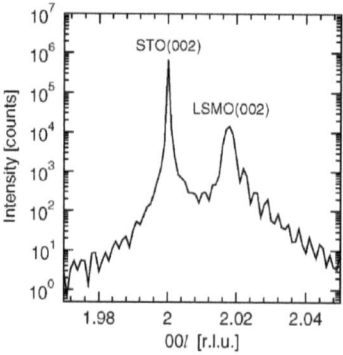

FIGURE 1 RHEED image of a 130 nm-thick $La_{0.66}Sr_{0.34}MnO_3$ thin film along the (100) azimuth. Note the 2-D zero-order and first-order Laue zone spots and the strong Kikuchi lines

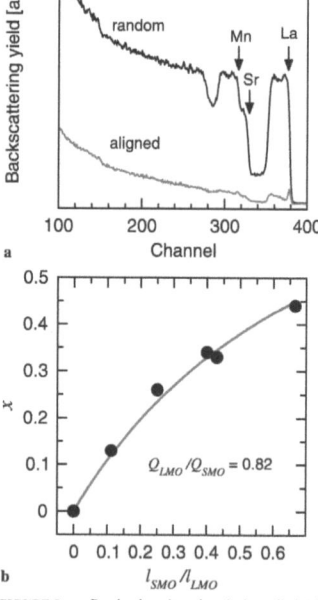

FIGURE 3 a Randomly oriented and channelled (aligned) RBS spectra of a 130 nm-thick $La_{0.66}Sr_{0.34}MnO_3$ thin film. b The change in film stoichiometry (x) of $La_{1-x}Sr_xMnO_3$ films as a function of the target length ratios l_{SMO}/l_{LMO}. The best-fit curve using (2) yields a deposition ratio of $Q_{LMO}/Q_{SMO} = 0.82$

FIGURE 2 $\theta - 2\theta$ XRD pattern of a 130 nm-thick $La_{0.66}Sr_{0.34}MnO_3$ thin film. The x-axis is given in reciprocal lattice units (r.l.u.) of the $SrTiO_3$ substrate system. Note the Kiessig fringes

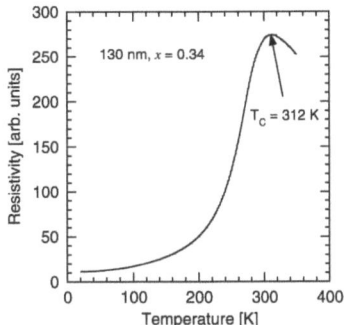

FIGURE 4 Four-point resistance measurements of a 130 nm-thick $La_{0.66}Sr_{0.34}MnO_3$ thin film as a function of temperature. The transition Curie temperature was 312 K

and can be fitted to a relative deposition yield of the components $Q_{LMO}/Q_{SMO} = 0.82$ [6] (Fig. 3b).

The film resistance as a function of temperature for $x = 0.34$ is shown in Fig. 4. It has a Curie temperature of 312 K, which lies some 50 K below the bulk value. This has been observed elsewhere [10–12] and is in agreement with finite-size scaling theory. Indeed, another film with a thickness of 106 nm showed a Curie temperature of 280 K.

4 Conclusions

$La_{1-x}Sr_xMnO_3$ thin films have been grown heteroepitaxially on $SrTiO_3$ using PRCLA and a novel ablation target geometry that allows any degree of doping between $x = 0$ and 1. The films were shown to be atomically flat, with a roughness of around 0.25 of a unit cell. Ion channelling demonstrated that the density of crystallographic defects was as low as that found in the $SrTiO_3$ substrate. The films should therefore be suitable for high-resolution ARPES measurements.

REFERENCES

1. S. Jin, T.H. Tiefel, M. McCormack, R.A. Fastnacht, R. Ramesh, L.H. Chen: Science **264**, 413 (1993)
2. J. Osterwalder: Surf. Rev. Lett. **4**, 391 (1997)
3. C. Zener: Phys. Rev. **82**, 403 (1951)
4. D.B. Chrisey, G.K. Hubler (Eds.): *Pulsed Laser Deposition of Thin Films* (Wiley, New York 1994)
5. P.R. Willmott, J.R. Huber: Rev. Mod. Phys. **72**, 315 (2000)
6. P.R. Willmott, R. Herger, C.M. Schlepütz: Thin Solid Films **453–454**, 436 (2004)
7. R. Timm, P.R. Willmott, J.R. Huber: Appl. Phys. Lett. **71**, 1966 (1997)
8. P. Lecoeur, A. Gupta, P.R. Duncombe, G.Q. Gong, G. Xiao: J. Appl. Phys. **80**, 513 (1996)
9. A. Gupta, B.W. Hussey: Appl. Phys. Lett. **58**, 1211 (1991)
10. M. Izumi, Y. Konishi, T. Nishihara, S. Hayashi, M. Shinohara, M. Kawasaki, Y. Tokura: Appl. Phys. Lett. **73**, 2497 (1998)
11. A.M. Haghiri-Gosnet, J. Wolfman, B. Mercey, C. Simon, P. Lecoeur, M. Korzenski, M. Hervieu, R. Desfeux, G. Baldinozzi: J. Appl. Phys. **88**, 4257 (2000)
12. M. Ziese, H.C. Semmelhack, K.H. Han, S.P. Sena, H.J. Blythe: J. Appl. Phys. **91**, 9930 (2002)

Paper IV
Energetic surface smoothing of complex metal-oxide thin films

The work presented in this chapter has been published in:
P.R. Willmott, R. Herger, C.M. Schlepütz, D. Martoccia, and B.D. Patterson, *Energetic surface smoothing of complex metal-oxide thin films*, Phys. Rev. Lett. **96**, 176102 (2006).

artikel4.pdf is the online version of the library (Zentralbibliothek Zürich).

Abstract

A novel energetic smoothing mechanism in the growth of complex metal oxide thin films is reported from *in situ* kinetic studies of pulsed laser deposition of $La_{1-x}Sr_xMnO_3$ on $SrTiO_3$, using x-ray reflectivity. Below 50% monolayer coverage, prompt insertion of energetic impinging species into small-diameter islands causes them to break up to form daughter islands. This smoothing mechanism therefore inhibits the formation of large-diameter 2D islands and the seeding of 3D growth. Above 50% coverage, islands begin to coalesce and their breakup is thereby suppressed. The energy of the incident flux is instead rechanneled into enhanced surface diffusion, which leads to an increase in the effective surface temperature of $\Delta T \approx 500$ K. These results have important implications on optimal conditions for nanoscale device fabrication using these materials.

DOI: 10.1103/PhysRevLett.96.176102 [a]
PACS numbers: 68.55.Ac, 61.10.Kw, 68.47.Gh, 81.15.Fg

Reprinted with kind permission from the American Physical Society.
[a] Note that you need a subscription for this journal to directly access the article.

Energetic Surface Smoothing of Complex Metal-Oxide Thin Films

P. R. Willmott,* R. Herger, C. M. Schlepütz, D. Martoccia, and B. D. Patterson

Swiss Light Source, Paul Scherrer Institut, CH-5232 Villigen, Switzerland
(Received 19 January 2006; published 2 May 2006)

A novel energetic smoothing mechanism in the growth of complex metal-oxide thin films is reported from *in situ* kinetic studies of pulsed laser deposition of $La_{1-x}Sr_xMnO_3$ on $SrTiO_3$, using x-ray reflectivity. Below 50% monolayer coverage, prompt insertion of energetic impinging species into small-diameter islands causes them to break up to form daughter islands. This smoothing mechanism therefore inhibits the formation of large-diameter 2D islands and the seeding of 3D growth. Above 50% coverage, islands begin to coalesce and their breakup is thereby suppressed. The energy of the incident flux is instead rechanneled into enhanced surface diffusion, which leads to an increase in the effective surface temperature of $\Delta T \approx 500$ K. These results have important implications on optimal conditions for nanoscale device fabrication using these materials.

DOI: 10.1103/PhysRevLett.96.176102   PACS numbers: 68.55.Ac, 61.10.Kw, 68.47.Gh, 81.15.Fg

One of the primary goals of modern condensed-matter physics is to facilitate the use in solid-state devices of novel materials, such as the diverse family of strongly correlated electron systems [1]. Their chemical and crystallographic complexity, however, presents a formidable challenge regarding the control of morphology and crystalline quality during film growth. In this respect, nonthermal growth techniques such as pulsed laser deposition (PLD) and sputtering have proved to be among the most promising to date [2,3]—the energetic particle beams used in these methods interact with the growing surface in ways that are unavailable to thermal deposition techniques [4–6]. This can lead to unusual and often advantageous growth kinetics and can force a film to grow under conditions far from thermal equilibrium [7–10]. A deeper understanding of the underlying atomistic processes is thus important for optimizing growth conditions to obtain nanoscale structures of high-quality material.

Because of its nonthermal nature, PLD is one of only a handful of techniques that is able to transfer chemically complex material congruently from the bulk to thin film [2]. In addition, the pulsed flux (of the order of 10^{20} atoms cm^{-2} s^{-1}) and associated supersaturation above the surface promotes the initial dense nucleation of small two-dimensional islands. These can be as small as a single atom and help promote two-dimensional film growth [3,11]. Before a monolayer (ML) is completed, however, conventional (i.e., thermal) growth models predict that the next monolayer(s) will begin to seed far away from step edges on large 2D islands, leading eventually to 3D growth [11]. How soon this happens depends on the effective surface diffusion constants (a) between the islands in the growing layer, leading to 2D island ripening, and (b) on top of the islands, resulting in interlayer mass transport [12]. This simple treatment is unable to explain the many reports in the literature in which thin films consisting of several hundred ML continue to grow two dimensionally, because it ignores processes involving the redistribution of the kinetic energy of the incoming flux as it impinges on the surface layer. It is these processes which need to be better understood.

Several theoretical studies have addressed this issue [4,5,13]. Particularly relevant is the work presented by Jacobsen *et al.*, in which the nonthermal interaction of an energetic incident atom with the surface (which occurs on the ps time scale) and thermal diffusion processes on the time scale of seconds were described by molecular dynamics (MD) simulations and kinetic Monte Carlo simulations, respectively, in a single model [14,15]. These studies of homoepitaxy of transition metals indicated that, as well as enhancing surface diffusion, particles impinging with kinetic energies of tens to hundreds of eV promoted smooth growth by inserting themselves into surface islands, provided they land near the descending edge of an island or terrace.

In this Letter, we report on a possible mechanism for promoting 2D growth in complex metal-oxide thin films, involving impact-induced island breakup, during PLD of the perovskite $La_{1-x}Sr_xMnO_3$ (LSMO) grown heteroepitaxially on $SrTiO_3$ (STO).

Experiments were performed at the Surface Diffraction station of the Materials Science beam line at the Swiss Light Source [16]. The *in situ* PLD setup and evidence for the high crystalline quality of the films have been described elsewhere [17,18]. Films of LSMO were grown on TiO_2-terminated STO(001) by ablating a dual $LaMnO_3$/$SrMnO_3$ target rod. In this study, $x = 0.34$ (checked by Rutherford backscattering spectroscopy). Film growth was monitored by recording the x-ray reflectivity at the (0 0 $\frac{1}{2}$) point of the specular crystal truncation rod (CTR). This signal oscillates during 2D island coalescence growth, due to (a) repeated roughening and smoothing, with a periodicity of 1 ML, and (b) interference between reflections from the film surface and the film-substrate interface (i.e., Kiessig fringes), with a periodicity of 2 ML.

Two deposition modes were employed. The first was "conventional" PLD growth at 10 Hz using a fixed repetition-rate Nd:YAG laser ($\lambda = 266$ nm). The 1 Hz frame rate of the pixel area x-ray detector employed for these studies therefore precluded the resolution of any thermal relaxation in between laser shots. To observe thermal kinetic effects, an interrupted mode was also used, in which short bursts of ablation (here, 12 laser shots, the minimum needed to ensure that the composition x of the film remained constant [17]) were separated from one another by intervals of a few tens of seconds of no deposition, during which changes in the specular intensity due to thermal relaxation could be monitored. This interrupted PLD is similar to pulsed laser *interval* deposition described by Rijnders et al., [8], the difference being that here only small fractions of a ML are deposited with each burst.

Growth oscillations during conventional PLD are shown in Fig. 1. Two-dimensional growth persists up to several tens of nm film thickness. Reflectivity curves of films over 100 nm thick show clear Kiessig fringes and can be fit to a roughness of well under 1 ML [17]. Surface smoothing therefore appears to be sufficiently rapid that little or no buildup occurs on top of each growing ML within the 20 s required for completion.

An example of growth using interrupted PLD is shown in Fig. 2 for deposition of the third ML. A description of the changes in the surface step density due to thermal surface diffusion has been proposed by Stoyanov and Michailov [12] and adapted to transient reflection high-energy electron-diffraction signals in PLD by Rijnders [19], although the model does not take into account changes in surface morphology due to energetic effects. Two constants τ_1 and τ_2 describe, respectively, the characteristic time to reach a step edge within the unfilled parts of the growing ML and that to cross the top of the islands making up the incomplete ML and drop into the growing layer. τ_1 dominates at low coverage θ, while τ_2 governs at higher coverages. For reasons that will become clear, it was difficult using the present temporal resolution to obtain reliable fits for τ_1, and we concentrate here on the second half of ML coverage, dictated by τ_2. In a first approximation, by ignoring τ_1, τ_2 can be fit by the expression [19]

$$I = I_0 + \Delta I\{1 - \exp[-(t - t_0)/\tau_2]\}, \quad (1)$$

whereby I_0 and t_0 are the intensity and time immediately after the 12-shot ablation burst, respectively, ΔI is the change in the specular reflection after complete thermal relaxation, and τ_2 is given by

$$\tau_2 = \frac{\theta}{D_S(\mu_l^{(0)})^2 \pi N_S}. \quad (2)$$

Here, $\mu_l^{(0)} = 2.40$ is the first root of the zeroth order Bessel function, N_S is the nucleation density, and D_S is the surface diffusion coefficient given by $D_S = D_0 \exp(-E_a/kT)$, where E_a is the surface diffusion barrier. Note that in this model, τ_2 is directly proportional to θ and should increase linearly from the start of the ML coverage.

The data for the second half of the ML growth have been fit using two free parameters ΔI and τ_2, and I_0 and t_0, which were allowed to deviate from their starting estimates only over ±250 arbitrary intensity units and ±1 s, respectively. The fits and the change in τ_2 with ML coverage θ are shown in the inset of Fig. 2. The thermal relaxation times are several tens of seconds for the second half of monolayer coverage, i.e., significantly longer than the time to grow a single ML when using conventional PLD. This difference demonstrates the beneficial influence of the impinging flux on the evolution and kinetics of the incomplete monolayer, which we will address more quantitatively below.

The high degree of supersaturation of the impinging flux of particles in PLD results in a high density of stable nucleation sites, each of the order of 1 to 2 atoms in size [3]. MD calculations predict an enhanced surface diffusion length of the order of 10 atomic spacings in the first few picoseconds after the particle lands on the surface [20]. The first pulse of impinging particles nucleates unit-cell or

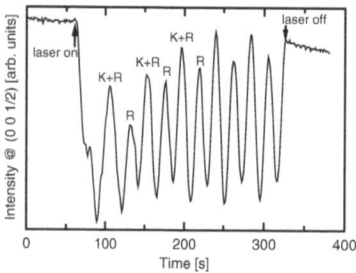

FIG. 1. Growth oscillations during conventional PLD of LSMO on STO(001). The signal at the $(00\frac{1}{2})$ position of the specular CTR is modulated by roughness fringes (R) with a periodicity of 1 ML and Kiessig fringes (K) every 2 MLs.

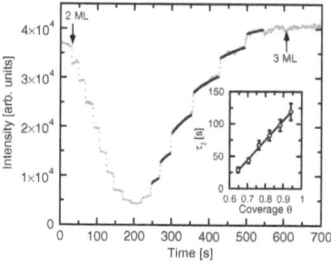

FIG. 2. Evolution of the $(0\,0\,\frac{1}{2})$ specular x-ray signal during the deposition of 1 ML of LSMO on STO, using 17 12-laser shot bursts, separated from one another by several seconds to allow the atoms to thermally diffuse to their optimal sites and minimize the atomic roughness. The change in x-ray reflectivity between laser bursts was fit using Eq. (1) and is shown by the solid black curves. The inset shows the time constants for these fits. A linear regression fit of the dependence of τ_2 on θ is also shown (straight solid line).

smaller sized islands, which are hence separated from one another approximately by a distance equal to the unit-cell size multiplied by the square root of the number of shots required to deposit a single ML (here $\sqrt{200} \approx 14$), i.e., 5 to 6 nm. This simple model does not consider the strongly ionic nature of the deposited material—the chemical potential to spontaneously produce ordered unit cells for these ionic materials is high and will assist adatoms to diffuse to optimal sites.

The diffusion constant τ_2 is predicted by Eq. (2) to increase linearly from the start of ML growth. A linear regression of the data in the inset of Fig. 2 shows, however, that τ_2 becomes nonzero only after $\theta \approx 0.55$. Although at very low coverages, hyperthermal surface diffusion on the picosecond time scale, induced by the impinging species, can also enhance the nucleation of the stable nucleation sites, prior to this critical coverage no significant thermal surface diffusion occurs across the top of the growing ML and the incident atoms are promptly inserted into the growing layer at the point of impingement. This can be explained as follows: The small island sizes imply that any atom impinging on top of them will be necessarily close to an island edge, allowing the prompt insertion mechanism for species having kinetic energies of the order of 10–20 eV. If the result is merely an increase in the lateral size of the island, however, this will enhance its resistance to any subsequent insertion events as lateral pushout of material becomes increasingly energetically unfavorable. Also, the probability of a particle impinging close to an island edge will become smaller as the islands spread laterally. However, theoretical studies for energetic Ag and Cu homoepitaxy have identified a second mechanism of island "breakup" or "chipping," which produces small daughter islands and suppresses the growth of large 2D islands [14,15]. Could this mechanism cause a similar island breakup in PLD of more strongly and directionally bonded ionic oxide systems?

To answer this, we consider three surface configurations which may be typical: (i) atoms embedded in the surface layer far from an island edge; (ii) five atoms (La/Sr, Mn, and 3 O atoms) making up a surface unit cell [Fig. 3(a)]; (iii) ten atoms consisting of two such units cells in a line [Fig. 3(b)]. We take tabulated bond strengths of La-O, Sr-O, and Mn-O of 800, 430, and 400 kJ mol^{-1} [21], and make the simplification of ignoring changes in individual dangling-bond energies. Surface La/Sr and Mn atoms not at island edges [i.e., case (i)] are eightfold and fivefold coordinated, respectively; hence transfer of kinetic energy to breaking bonds in excess of 20, 25, and 65 eV is required for sputtering of Mn, Sr, and La, respectively [22]. This will be most efficient if the impinging atom has the same mass as the struck surface atom. Typical kinetic energies for ablation species in PLD lie in the range of 5 to 25 eV [2]. Therefore disruption of surface atoms not at an island edge appears to be inefficient.

An isolated surface unit cell of La$_{1-x}$Sr$_x$MnO$_3$ [case (ii)] contains 3 La/Sr-O and 3 Mn-O bonds (not counting those

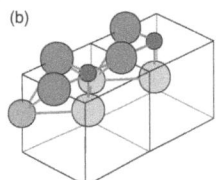

FIG. 3 (color). The positions and bonds of (a) a single unit-cell island and (b) a double unit-cell island. La/Sr, green; Mn, red; O, blue; La/Sr-O bond, cyan; Mn-O bond, magenta.

connecting the unit cell perpendicularly to the substrate), while an island consisting of two unit cells [i.e., case (iii)] contains 8 La/Sr-O and 7 Mn-O bonds. The difference in bonding energy between a double unit-cell island and 2 single unit-cell islands is therefore ≈ 16 eV, well within the typical energy range of the impinging species. The activation barrier between these two configurations is more difficult to estimate, but is likely to be reduced by electron-phonon coupling on the picosecond time scale [23].

Assuming the validity of the island breakup mechanism, the density of small islands will increase until they begin to coalesce, which will occur when the distance between them is approximately the same as their size, i.e., at $\theta = 0.5$ [19], which indeed we observe. Prior to this, there is negligible τ_2 diffusion; hence, Eq. (2) is modified to

$$\tau_2 = \frac{\theta - \theta_{min}}{D_S(\mu_l^{(0)})^2 \pi N_S}, \qquad (3)$$

where θ_{min} is the minimum coverage, below which mass transfer from on top of the growing layer to the growing layer is via prompt insertion. After this minimum coverage is reached, nonthermal smoothing in conventional PLD continues not only because the impinging particles have a transiently enhanced surface diffusion, but also because their transferred energy couples to those adatoms on the surface which have yet to find a stable site. This model appears to be particularly strong compared to other known competing nonthermal mechanisms (such as resputtering, enhanced surface diffusion alone, or implantation) as it is the only one that would show the sudden change in relaxation behavior at $\approx 50\%$ ML coverage, due to the fact it is the only one that depends on the microscopic morphology (i.e., the islands' size and surface density).

Therefore, for the first half of ML coverage, nucleation and island breakup at or near the point of impingement dominate. The average distance for adatoms *thermally* diffusing in the growing ML to ascending island edges (which determines τ_1) is for low θ comparable to the enhanced diffusion length and thereafter becomes smaller, which explains why τ_1 is so much shorter than τ_2. The processes leading to 2D monolayer growth in this model are summarized schematically in Fig. 4.

We can estimate the effective increase in surface temperature induced by PLD from a comparison of times to

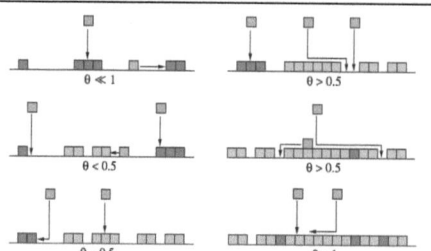

FIG. 4 (color). Summary of the proposed processes influencing long-term 2D film growth by PLD. At low coverage, the impinging particles (red) nucleate small, relatively densely packed islands (blue). Any impinging particles landing on top of these islands can cause them to split into daughter islands (green). The density of the small islands subsequently increases, such that at ≈ 50% coverage, their separation is approximately equal to their size, and the islands begin to coalesce, excluding further breakup. After this, the impinging species must diffuse to descending edge steps, which is accelerated by the energy of the impinging particles now coupling to diffusing adatoms.

produce atomically flat layers in conventional PLD and the characteristic thermal diffusion times between laser bursts in interrupted PLD. If we assume that typical surface diffusion constants for metal oxides lie between $D_0 = 10^{-4} \rightarrow 10^{-8}$ cm^2 s^{-1}, and that, toward ML completion, the adatoms have to travel between $\rho = 4$ and 25 nm to find a stable site, we obtain from the thermal diffusion data in the inset of Fig. 2 and a growth temperature of 1000 K a value of E_a that ranges between 2.2 eV ($D_0 = 10^{-4}$ cm^2 s^{-1}, $\rho = 4$ nm) and 1.1 eV ($D_0 = 10^{-8}$ cm^2 s^{-1}, $\rho = 25$ nm). We have seen in conventional PLD that enhancement of surface diffusion enables all the adatoms to find stable sites within the 0.1 s in between laser shots, even at high coverage. Using our values for E_a, we obtain an effective increase in the surface temperature of between 370 K ($D_0 = 10^{-4}$ cm^2 s^{-1}, $\rho = 4$ nm) and 1200 K ($D_0 = 10^{-8}$ cm^2 s^{-1}, $\rho = 25$ nm). These represent the limits. "Typical" estimates of $D_0 = 10^{-6}$ cm^2 and $\rho = 8$ nm yield $\Delta T \approx 500$ K.

In conclusion, prompt insertion of impinging ablation species into small-diameter 2D islands causing them to break up into daughter islands is proposed to explain the kinetics of long-term 2D thin film growth of complex metal-oxides using PLD. This mechanism restricts the lateral size of the islands up to a coverage of about 50%, at which point they begin to coalesce, shutting off the breakup mechanism. Thereafter, the kinetic energy of the impinging species is channeled into enhanced surface diffusion of the adatoms, which yields an effective increase in temperature (at least with respect to the degrees of freedom of lateral surface movement) of the order of 500 K. In this manner, monolayer completion is sufficiently efficient to allow 2D growth up to several tens of nm thickness. The authors believe such behavior has been recorded but not recognized elsewhere [8], and that a judicious tuning of the energies of the incident species (which can be brought about by adjusting the laser fluence and/or the degree of quenching by a background moderating gas) could lead to this becoming a general phenomenon for a large range of chemical systems. Finally, theoretical models of multielemental thin film growth far from thermal equilibrium can now be verified against these unique quantitative experimental results.

Support of this work by the Schweizerischer Nationalfonds zur Förderung der wissenschaftlichen Forschung and the staff of the Swiss Light Source is gratefully acknowledged.

*Electronic address: philip.willmott@psi.ch
[1] M. B. Salamon and M. Jaime, Rev. Mod. Phys. **73**, 583 (2001).
[2] P. R. Willmott and J. R. Huber, Rev. Mod. Phys. **72**, 315 (2000).
[3] P. R. Willmott, Prog. Surf. Sci. **76**, 163 (2004).
[4] G. K. Hubler and J. A. Sprague, Surf. Coat. Technol. **81**, 29 (1996).
[5] M. E. Taylor and H. A. Atwater, Appl. Surf. Sci. **127–129**, 159 (1998).
[6] R. M. Tromp and J. B. Hannon, Surf. Rev. Lett. **9**, 1565 (2002).
[7] H. Jenniches, M. Klaua, H. Höche, and J. Kirschner, Appl. Phys. Lett. **69**, 3339 (1996).
[8] G. Rijnders, G. Koster, V. Leca, D. H. A. Blank, and H. Rogalla, Appl. Surf. Sci. **168**, 223 (2000).
[9] J. M. Warrender and M. J. Aziz, Appl. Phys. A **79**, 713 (2004).
[10] B. Shin, J. P. Leonard, J. W. McCamy, and M. J. Aziz, Appl. Phys. Lett. **87**, 181916 (2005).
[11] A. Fleet, D. Dale, Y. Suzuki, and J. D. Brock, Phys. Rev. Lett. **94**, 036102 (2005).
[12] S. Stoyanov and M. Michailov, Surf. Sci. **202**, 109 (1988).
[13] M. Villarba and H. Jónsson, Surf. Sci. **324**, 35 (1995).
[14] J. Jacobsen, B. H. Cooper, and J. P. Sethna, Phys. Rev. B **58**, 15847 (1998).
[15] J. M. Pomeroy, J. Jacobsen, C. C. Hill, B. H. Cooper, and J. P. Sethna, Phys. Rev. B **66**, 235412 (2002).
[16] B. D. Patterson et al., Nucl. Instrum. Methods Phys. Res., Sect. A **540**, 42 (2005).
[17] P. R. Willmott, R. Herger, M. C. Falub, L. Patthey, M. Döbeli, C. V. Falub, M. Shi, and M. Schneider, Appl. Phys. A **79**, 1199 (2004).
[18] P. R. Willmott et al., Appl. Surf. Sci. **247**, 188 (2005).
[19] A. J. H. M. Rijnders, Ph.D. thesis, University of Twente, 2001.
[20] T. Diaz de la Rubia, R. S. Averbach, R. Benedek, and W. E. King, Phys. Rev. Lett. **59**, 1930 (1987).
[21] *Handbook of Chemistry and Physics*, edited by D. R. Lide (CRC Press, Boca Raton, 1993).
[22] D. K. Brice, J. Y. Tsao, and S. T. Picraux, Nucl. Instrum. Methods Phys. Res., Sect. B **44**, 68 (1989).
[23] R. J. Hamers, Surf. Sci. **583**, 1 (2005).

Paper V

Structure determination of monolayer-by-monolayer grown $La_{1-x}Sr_xMnO_3$ thin films and the onset of magnetoresistance

The work presented in this chapter has been published in:
R. Herger, P.R. Willmott, C.M. Schlepütz, M. Björck, S.A. Pauli, D. Martoccia, B.D. Patterson, D. Kumah, R. Clarke, Y. Yacoby, and M. Döbeli, *Structure determination of monolayer-by-monolayer grown $La_{1-x}Sr_xMnO_3$ thin films and the onset of magnetoresistance*, Phys. Rev. B **77**, 084501 (2008).

artikel5.pdf is the online version of the library (Zentralbibliothek Zürich).

Abstract

Surface x-ray diffraction was used to determine the atomic structures of $La_{1-x}Sr_xMnO_3$ thin films, grown monolayer by monolayer on $SrTiO_3$ by pulsed laser deposition. Structures for one-, two-, three-, four-, six-, and nine-monolayer-thick films were solved using the Coherent Bragg rod analysis phase-retrieval method and subsequent structural refinement. Four important results were found. First, the out-of-plane lattice constant is elongated across the substrate-film interface. Second, the transition from substrate to film is not abrupt, but proceeds gradually over approximately three unit cells. Third, Sr segregates towards the topmost monolayer of the film: we determined a Sr-segregation enthalpy of -15 kJ/mol from the occupation parameters. Finally, the electronic bandwidth W was used to explain the onset of magnetoresistance for films of nine or more monolayers thickness. Resistivity measurements of the nine monolayer-thick film confirm magnetoresistance and the presence of a dead layer with mostly insulating properties.

DOI: 10.1103/PhysRevB.77.085401 [a]

PACS numbers: 68.47.Gh, 61.05.cp, 81.15.Fg, 75.47.Gk

Reprinted with kind permission from the American Physical Society.
[a] Note that you need a subscription for this journal to directly access the article.

Structure determination of monolayer-by-monolayer grown $La_{1-x}Sr_xMnO_3$ thin films and the onset of magnetoresistance

R. Herger, P. R. Willmott,* C. M. Schlepütz, M. Björck, S. A. Pauli, D. Martoccia, and B. D. Patterson
Swiss Light Source, Paul Scherrer Institut, CH-5232 Villigen, Switzerland

D. Kumah and R. Clarke
Randall Laboratory of Physics and FOCUS Center, University of Michigan, Ann Arbor, Michigan 48109-1120, USA

Y. Yacoby
Racah Institute of Physics, Hebrew University, Jerusalem 91904, Israel

M. Döbeli
Ion Beam Physics, Paul Scherrer Institut and ETH-Zurich, CH-8093 Zurich, Switzerland
(Received 28 September 2007; published 1 February 2008)

Surface x-ray diffraction was used to determine the atomic structures of $La_{1-x}Sr_xMnO_3$ thin films, grown monolayer by monolayer on $SrTiO_3$ by pulsed laser deposition. Structures for one-, two-, three-, four-, six-, and nine-monolayer-thick films were solved using the Coherent Bragg rod analysis phase-retrieval method and subsequent structural refinement. Four important results were found. First, the out-of-plane lattice constant is elongated across the substrate-film interface. Second, the transition from substrate to film is not abrupt, but proceeds gradually over approximately three unit cells. Third, Sr segregates towards the topmost monolayer of the film: we determined a Sr-segregation enthalpy of −15 kJ/mol from the occupation parameters. Finally, the electronic bandwidth W was used to explain the onset of magnetoresistance for films of nine or more monolayers thickness. Resistivity measurements of the nine monolayer-thick film confirm magnetoresistance and the presence of a dead layer with mostly insulating properties.

DOI: 10.1103/PhysRevB.77.085401
PACS number(s): 68.47.Gh, 61.05.cp, 81.15.Fg, 75.47.Gk

I. INTRODUCTION

Ferromagnetic manganites were first investigated by Jonker and van Santen in 1950, but have attracted renewed interest in recent years since the discovery that they exhibit colossal magnetoresistance (CMR).[1,2] Doped manganites with the perovskite structure and chemical composition $RE_{1-x}AE_xMnO_3$, where RE is a rare earth and AE is a divalent alkaline earth, show rich phase diagrams, due to the complex interplay of charge, spin, lattice, and orbital degrees of freedom.[3,4] Their interesting physical properties have not only triggered renewed scientific interest in these compounds, but also show potential for many technological applications such as spin electronics or magnetic sensors. Thin films are best suited for these demands.

The ongoing trend of miniaturization means that novel materials in the form of thin films are very important for any technological application. Surface and interface effects can set a lower limit to downsizing devices that exploit bulk effects. These mainly structural considerations become particularly important in ultrathin films, where surface and interface relaxations can involve a significant fraction of the film volume and hence fundamentally change the physical properties. An exact knowledge of the atomic positions is therefore of great importance for the design of nanoscale devices.

Bulk $La_{1-x}Sr_xMnO_3$ (LSMO) at an optimal doping of $x = 1/3$ shows the transition from a paramagnetic insulator to a ferromagnetic metal at temperatures as high as $T_C = 370$ K.[4] The Mn site has a mixed valence state of x Mn^{4+} (holes) and $(1-x)$ Mn^{3+}, leading to degenerate high-spin t_{2g}^3 and $t_{2g}^3 e_g^1$ states of the MnO_6 octahedra, respectively, due to the large Hund exchange energy of $J_H \approx 2.5$ eV in the crystal field. The electrons can hop between adjacent Mn ions, as described by Zener's double-exchange mechanism[5] and thus mediate the long-range ferromagnetic ground state of the metallic conductor. Additionally, as the electronic ground state of the Mn^{3+} sites is degenerate, a Jahn-Teller distortion breaks the octahedral symmetry and lowers the energy.[6]

In thin films, however, the transport properties of LSMO change dramatically. The typical shape of the metal-insulator transition of the bulk changes to a more semiconductorlike behavior of the resistivity curve in ultrathin films, with resistivities in the low temperature regime being approximately four orders of magnitude higher. This has been explained by the presence of an electrical "dead layer" at the substrate-film interface.[7–9] Nevertheless, such films still exhibit magnetoresistive behavior.[7] Our intention is thus to use exact atomic coordinates to correlate the structure with the transport properties, in order to determine a minimum thickness for the onset of magnetoresistance.

In this work, we present a detailed structure determination via surface x-ray diffraction (SXRD) of thin LSMO films of 1-, 2-, 3-, 4-, 6-, and 9-monolayer (ML) thickness, grown by pulsed laser deposition (PLD) pseudomorphically on (001) $SrTiO_3$ (STO, cubic lattice constant $a_{STO} = 3.905$ Å). A structural study of different film thicknesses has enabled us to monitor the evolution of the growth of thin films with sub-Å resolution, revealing interesting structural features.

The main results of the structure determination are (i) the observation of a dilation of the interface perpendicular to the surface, in contrast to the compression that one would intuitively expect due to the tensile in-plane stress of LSMO [quasicubic lattice constant of $a_{LSMO}=3.875$ Å ($x=0.35$) (Ref. 4)] grown on STO; (ii) the formation of a nonabrupt interface that is used to derive a general picture of how the film stoichiometry becomes established; (iii) the enrichment of Sr in the topmost layer of the film surface with an approximate estimate of the segregation enthalpy based on occupation parameters; and (iv) the fact that magnetoresistance can be observed in films with nine or more ML thickness, explained by considerations of the electronic bandwidth using the structural data and by comparison with the bulk properties.

II. METHODS

A. Experiment

1. Pulsed laser deposition

LSMO thin films were grown in an *in situ* PLD chamber mounted on a surface x-ray diffractometer.[10] The films were deposited on STO(001) substrates with low vicinality ($<0.1°$) prepared by an established chemical and thermal treatment to ensure TiO_2 termination.[11,12]

We used the fourth harmonic of a Nd:YAG laser (10 Hz, 266 nm, 5 ns, 2 J cm^{-2}) as the ablation source, in conjunction with a synchronized N_2O gas pulse (1.5×10^{-2} Pa, average pressure) and an O_2 background (2×10^{-2} Pa). These conditions led to films of high crystallinity.[13] The laser was operated in an interrupted mode, in which short bursts of ablation (typically 12 shots, the minimum required to ensure the conservation of the chosen stoichiometry of $x=0.35$) are separated by a period of several tens of seconds, in order to let the surface thermally relax. After deposition, the films were quenched in oxygen.

Here, we report on films that are 1, 2, 3, 4, 6, and 9 MLs thick. Henceforth, we refer to a film of y ML thickness (i.e., the substrate surface covered by an LSMO layer of exactly y unit cells in height) as LSMO y, e.g., the 4-ML-thick film is LSMO 4. The films were grown on three different STO substrates 1, 2, and 3, as described in Table I.

2. Surface x-ray diffraction

Surface x-ray diffraction experiments were carried out at the surface diffraction station of the Materials Science Beamline at the Swiss Light Source, Paul Scherrer Institut.[15] The growth chamber contains a large beryllium window,[10] enabling *in situ* data acquisition. The (2+3) surface diffractometer is equipped with a fast, single photon-counting 2D x-ray pixel detector. We used 1-Å synchrotron radiation and a fixed incidence angle of 0.15°, slightly below the critical angle of 0.20°, in order to enhance the surface signal.

After every deposition, we recorded a large SXRD data set, typically consisting of ten inequivalent and about five equivalent CTRs (see Table I for details). From the pixel images, the integrated intensities were extracted and standard

TABLE I. Details of SXRD data sets and refinement parameters. Film thicknesses are given in MLs and STO denotes the used substrate. The number of inequivalent (ineq.) and equivalent (eq.) structure factors (SFs) are given with the number of measured CTRs in brackets. ε is the systematic error and P is the number of fit parameters used with the resulting oversampling factor (O). The final R factor $R(|F|)$ after structure refinement is given. The surface occupation (SO) is in fractions of a ML.

Film	STO	SFs (CTRs) Ineq.	Eq.	ε (%)	P	O	$R(|F|)$ (%)	SO
1	1	369(10)	211(5)	17.1	36	10.3	13.3	0.897
2	1	408(11)	316(9)	9.6	48	8.5	8.2	0.920
3	1	369(10)	157(4)	5.3	60	6.2	8.3	0.891
4	2	418(11)	303(9)	16.4	72	5.8	16.4	0.925
6	3	375(10)	315(6)	11.2	96	3.9	10.3	0.999
9	3	1029(10)	0(0)	11.2^a	132	7.8	10.2	1.005

^{a}Due to the lack of equivalent reflections, ε was assumed to be the same as for LSMO 6 on the same STO sample 3.

geometrical correction factors applied.[16] Thus we obtained about 400 nonequivalent structure factors for each thickness, resulting in systematic errors ranging from 5 to 17 %, primarily attributable to mechanical distortions (e.g., bending of the substrate) produced by the heater/clamping mechanism.

The data span reciprocal space in $|h|$ and $|k|$ from 0 to 4. Perpendicular to the surface, we selected reflections with $0.5 \leq l < 3$, in order to account for the sampling resolution along l, the quality of the STO substrates, and the angle of incidence, as reasoned elsewhere.[17] None of the films showed reconstructions.

3. Ex situ characterization

(a) *Rutherford backscattering*. Rutherford backscattering (RBS) experiments were carried out using a 2-MeV ^{4}He beam and a silicon surface barrier detector at a scattering angle of 165°.[18] The background was subtracted using a recently developed fitting procedure.[19] For thicker films of the order several tens of nm (not presented here) grown using the same conditions as described in Sec. II A 1, the elemental composition could be determined using the RUMP program.[20] For very thin films consisting of only a few MLs, however, the RBS analysis for Sr and O fails. Nevertheless, the stoichiometry could be obtained by element-specific integration of the backscattering signal of La and Mn and their correlation to the integrated signals of the thicker films, under the assumption that the backscattering yield for thin films is proportional to that for thick, more bulklike films.

In addition, RBS channeling experiments yielded information on the growth quality and crystallographic defects such as dislocations or interface roughness. A representative channeling spectrum of a 130-nm-thick LSMO film can be found in Ref. 13.

(b) *X-ray reflectivity*. We recorded x-ray reflectivity (XRR) curves at a wavelength of 1 Å in order to determine the thickness of LSMO 6 and LSMO 9. The reflected signal for each incident angle was integrated and corrected for the footprint of the x-ray beam on the sample surface. The reflectivity curves were fit using the program GENX.[21]

(c) *X-ray diffraction.* Laboratory-based $\theta-2\theta$ scans using Cu $K\alpha$ radiation confirmed single crystal growth along the (001) axis.

(d) *Resistivity.* The resistivity was measured using the four-point method in a customized sample holder. We used a Quantum Design model 6000 physical properties measurement system (PPMS) to measure the electrical resistance R without and with an applied field of $B=5$ T perpendicular to the film surface. Every point was measured 25 times, and the standard deviation was calculated. Before the resistivity measurements, LSMO 3, 4, and 9 were annealed in pure, flowing oxygen at 900 °C for 3 h in order to enhance the transport properties of the thin film.

B. Structural analysis

Coherent Bragg rod analysis (COBRA) is a direct phase-retrieval algorithm for SXRD data.[22] The COBRA method is generally applicable to systems that are periodic in two dimensions, aperiodic in the third, and commensurate with the underlying substrate. COBRA provides a 3D electron density map of the system with sub-Å resolution. For each film thickness, a reference structure consisting of an undistorted film and substrate with bulk positions and nominal thickness and composition was used. Convergence was achieved after three to four small COBRA phasing iterations.[22,23] In Fig. 1, the COBRA result for LSMO 4 is presented as an example.

To obtain more precise occupancy results, the atomic positions and the occupancies of the film layers determined by COBRA were used as a starting model for subsequent structure refinement in the FIT program.[24] Structural refinement robustly converged for a given film thickness. All thicknesses were modeled using consistent conditions (i.e., concerning symmetry, fit parameters, etc.) and thus the refined models enable one to directly compare the results. Optimization was carried out by minimizing the crystallographic R factor.[25] It is emphasized that without the initial COBRA phasing, there are too many unknown parameters for the fitting and refinement approach alone to converge on a viable solution.

The structures were modeled as follows. Atoms were allowed to move only in the z direction, i.e., perpendicular to the surface, according to the $p4mm$ surface symmetry. The positive z direction (in units of bulk STO unit cells) was defined as pointing out of the surface. The positions of all the film atoms and the three top MLs of bulk STO ($z>-3$, with the top substrate TiO$_2$ layer as the nominal zero position) were refined. It was found that fitting the positions of La independently from that of Sr on the same site, and Ti independently of Mn, made no significant improvements to the fit. Hence La and Sr, as well as Ti and Mn, each used a common position parameter. Every atom from $z>-1$ up to the surface was assigned an individual isotropic Debye-Waller (DW) factor. For $z \leq -1$, the sites of each atom type had element-specific isotropic DW factors for Sr, Ti, and O. The occupation parameters of La and Sr, and Ti and Mn were refined with the restriction that the total occupation per site

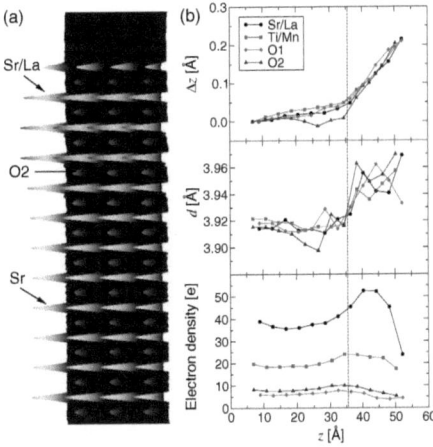

FIG. 1. (Color online) COBRA results for LSMO 4. (a) The electron density (ED) map obtained by the COBRA phasing method, showing the plane along z containing the La, Sr, and O2 atoms. (b) The cumulative displacement Δz of the atoms from the reference frame of bulk STO (top), the distances d between neighboring atoms of the same element across the substrate-film interface (middle), and the integrated electron densities of the Gaussian-like features in the ED maps (bottom). Uncertainties in Δz and d are estimated to be ± 0.03 Å, while the electron densities are accurate to $\pm 5\%$.

had to be unity. We modeled an incomplete (or "overcomplete") ML coverage that might occur in the PLD growth process by allowing the top *monolayer*[26] of the film to have a noninteger occupation [see the surface occupation (SO) column in Table I]. This parameter also accounts for possible surface roughness, although the root mean square roughness is known to be only 1 to 2 Å for films of several tens of nanometers thickness.[7,27]

III. RESULTS AND DISCUSSION

A. Film characterization

1. Thickness of the films

For all LSMO films, the thickness was apparent from the growth oscillations. For the thicker films, however, we additionally used XRR to verify the thickness. Fitting with GenX gave 5.6(2) and 9.1(2) MLs for LSMO 6 and LSMO 9, respectively, using the bulk LSMO lattice constant. The XRR fits also reveal that the films consist of an approximately 22-Å-thick interface region of higher optical density and, for LSMO 9, a second, graded, less dense surface layer (up to 20% less at the surface)[28] of approximately 10 Å depth, which is probably caused by surface roughness and/or incomplete monolayer coverage. The XRR result for LSMO 6 is not in perfect agreement with the nominal 6 MLs. From

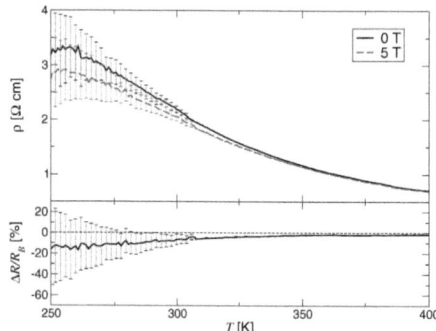

FIG. 2. (Color online) Top: Resistivity curves of LSMO 9 without magnetic field (solid black) and at $B=5$ T (dashed red). Bottom: Magnetoresistance ratio $\Delta R/R_B = (R_B - R_0)/R_B$. The dashed line is a guide to the eye and indicates zero magnetoresistance. Only every second error bar is drawn to improve the readability.

Table I and the growth oscillations, we have to conclude that this film has a thickness very close to 6 MLs. This is also supported by the fact that LSMO 9 is the continuation of the growth of LSMO 6.

2. Resistivity measurements

Experimental resistivity curves of LSMO 9 are shown in Fig. 2.[29] The resistivity ρ is relatively high, as would be expected for a film of only 9 ML thickness, and shows semiconducting behavior. Applying a field of $B=5$ T reduces ρ, i.e., we see a magnetoresistive effect. Note that the error bars (i.e., the standard deviations) below 310 K suddenly increase. We attribute this to the voltage limit of the PPMS. We get a negative magnetoresistance ratio $\Delta R/R_B = (R_B - R_0)/R_B = -7\%$ at 310 K. This value is far from being "colossal," but consistent with other experimental observations for comparably thin films.[7,8]

The curves do not show the usual metal-insulator transition that one might expect for thicker, bulklike films. Instead, they resemble more semiconducting behavior, indicating the presence of an electrical dead layer at the substrate-film interface. Recent theoretical investigations reveal this to be an intrinsic phenomenon at a metal-insulator interface.[30] These dead layers were found all to have similar thicknesses, independent of the substrate used: 5 (12), 5 (12), 3 (8), and 4 (10) nm (MLs) for LSMO grown on STO, LaAlO$_3$, NdGaO$_3$, and MgO, respectively.[7–9] Interestingly, Sun et al.[8] ruled out that the presence of these dead layers is caused by strain and concluded that the substrate-film interface and/or the surface must be chemically or structurally altered in LSMO thin films. Liao et al.[7] gave a possible explanation for the simultaneous presence of magnetoresistance and high resistivity. From the observation of a spatially inhomogeneous metal-insulator transition, they concluded that phase separation leads to ferromagnetic metallic domains embedded in an insulating framework of the dead layer.

B. Structure

We present three selected CTRs [(11l), (22l), and (32l)] for each thickness plus the fit intensities in Fig. 3. The R factors range from 8.2 to 16.3 % (see Table I). The atomic coordinates and occupation parameters of the films are available online.[31] The discussion in the following sections is based on these data and will lead to the four main results of this work.

The surfaces of our thin films appear to have a low defect density, judged from inspection of the DW factors. Moreover, high-quality growth on STO can be achieved even for

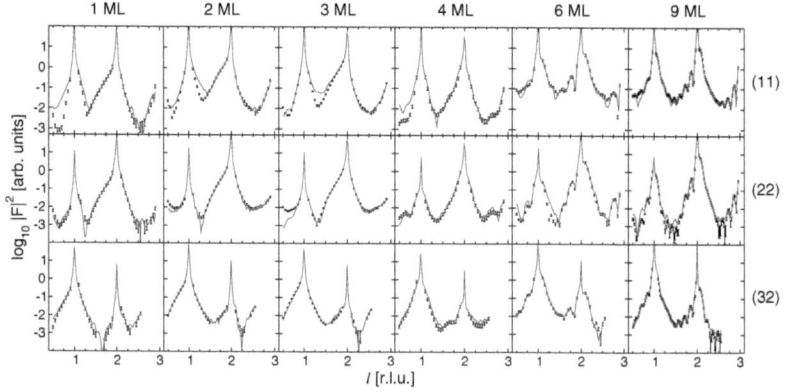

FIG. 3. (Color online) Sets of SXRD data (black) and calculated intensities (red) for different film thicknesses as labeled. For representation, three rods of the data files were selected: (11l) top, (22l) middle, and (32l) bottom.

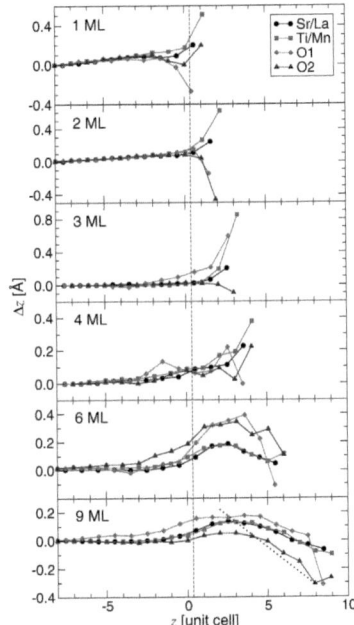

FIG. 4. (Color online) The cumulative displacements Δz of the atoms from the reference frame of the positions for bulk STO. The zero position in z represents the top TiO_2 layer of bulk STO with the nominal interface indicated by the dashed line. Color and symbol code: Sr/La=black ●, Ti/Mn=red ■, O1=green ◆, O2=blue ▲. All film thicknesses show an elongation of the out-of-plane lattice constant at the interface. Results for metallic sites are estimated to be accurate to ±0.03 Å, with the caveat that the positions in the topmost ML of the thinnest films (1–3 MLs) are likely to have significantly larger uncertainties, as are the oxygen displacements. Thinner films exhibit a more pronounced increase in the lattice constant than thicker films. The dotted line for LSMO 9 has a slope of $\Delta z = -0.09z + 0.4$ and indicates the LSMO lattice constant for strained bulk above $z \geq 3$ MLs, as detailed in the text.

thick LSMO films (i.e., several tens of nanometers), as can be seen from our RBS channeling results and the clear Kickuchi lines in the reflection high-energy electron-diffraction pattern[13] or by atomic force microscopy.[7,27]

C. Substrate-film interface

One of the main results of this work is the unusual behavior of the substrate-film interface. Tensile stress in-plane of the smaller LSMO unit cell grown on STO would, on its own, lead to a *decrease* in the out-of-plane lattice constant c. We see, however, the opposite phenomenon (Fig. 4): All our films show an *increase* in the out-of-plane lattice constant across the interface, with a maximum deviation from bulk STO after about 3 MLs. For films thicker than 3 MLs, this dilatory layer is capped by a layer where c decreases again. If we were to assume a *strained* bulk lattice constant for films thicker than 3 MLs ($z \geq 3$), the slope would be given by −0.09 Å. This is represented by the dotted line in Fig. 4. We note that $c > c_{LSMO,strained}$, i.e., the film is less strained than expected from this simple model taking only in-plane compressive strain into account.

Figure 4 also gives information on the reliability of the atomic positions, as the six films were grown on three different substrates. For the metallic sites, the positions are reproduced to within ±0.03 Å. The results for oxygen, with its low x-ray scattering power are likely to be significantly less reliable. Note also that the cumulative displacement in the very uppermost monolayer (i.e., at the growth surface) of the thinnest films (1–3 MLs) may be less reliable because of surface roughness and/or incomplete occupancy.

An important issue to be addressed is interfacial roughness. The refined DW factors at the interface are comparable to the tabulated bulk values,[32] indicating low uncertainties in the atomic positions. Sometimes, we note increased DW factors for Sr in the film. A possible explanation for this could be the preference of Sr to segregate, as discussed in Sec. III E. Our RBS channeling results for thicker films (≈ 130 nm) reveal almost no interfacial crystallographic defects.[13] We can therefore conclude that the interface roughness is low and plays a negligible role for any explanation of the behavior of the film-substrate interface.

As mentioned above, elongation of c at the interface cannot be explained by strain alone. A more likely explanation could involve the presence of ions in lower oxidation states. Such atoms have larger ionic radii and could account for the observed dilation. We will address this in the next section in more detail.

Finally, fitting of the (002) Bragg peak of STO from laboratory-based θ-2θ scans using Voigt line profiles corroborated the increase in the out-of-plane lattice constant in LSMO 3, 4, and 9. Only LSMO 9 additionally showed a significant signal attributable to a c smaller than c_{STO}, which probably can be associated with the upper MLs of the film.

D. Stoichiometry

We present the change in occupancies for the metallic sites across the substrate-film interface in Fig. 5. The transition from substrate to film is not abrupt, but proceeds gradually over two to three MLs and already starts in the nominally top ML of the STO substrate.

The COBRA results give clear evidence for the gradually changing interface (Fig. 1). It is unlikely that this effect is merely an artefact of the COBRA analysis, based on the following arguments. Ultra thin two-dimensional (2D)-grown films of the order of several MLs tend to mimic the substrate surface, i.e., flat substrate surfaces will translate to flat film surfaces. The COBRA results reveal such a behavior: The integrated electron density (ED) drops off very sharply within half a ML at the film surface, suggesting a flat film surface and a (initially) flat substrate surface. However, the ED gradually increases over approximately 3 MLs, sug-

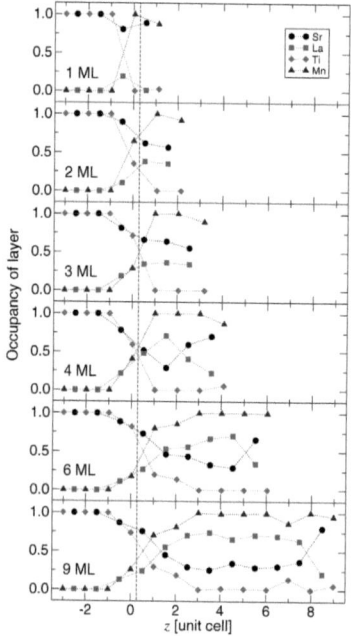

FIG. 5. (Color online) Occupancies across the substrate-film interface (dashed line) for different film thicknesses. The change in stoichiometry takes place over approximately three unit cells. Sr =black •, La=red ■, Ti=green ♦, Mn=blue ▲.

TABLE II. Average occupation parameters determined with FIT compared with RBS results for La and Sr.

LSMO	FIT		RBS	
	La	Sr	La	Sr
3	0.43	0.57	0.69	0.31
4	0.52	0.48	0.64	0.36
9	0.59	0.41	0.62	0.38

as can be seen in Fig. 5. The transition from Ti to Mn seems to indicate an enrichment of Mn, whereby the number of MnO_2 layers exceeds the nominal film thickness in MLs by 1, as can be seen by the Mn-rich ML of LSMO 1 below the nominal interface. However, it remains unclear where this extra Mn should come from. A more likely explanation for this could be the presence of additional Ti in the interface region, in combination with the fact of similar scattering behavior of Mn and Ti when probed with x rays. In Ref. 12 we found the STO substrate under thin film growth conditions to be terminated with a double TiO_2 layer. Moreover, growth oscillations show a peculiar behavior immediately after starting deposition, which could be interpreted as a fingerprint of the incorporation of the TiO_2-double layer into the film structure (see, for example, Fig. 1 in Ref. 14). From the COBRA results, we do not see that a TiO_2-double layer, as proposed in Ref. 12, is present at the substrate-film interface. But the integration of the additional TiO_2-layer material into the film structure remains an interesting suggestion. More work on the growth modes of LSMO thin films has to be carried out to resolve this issue unambiguously.

We are now able to summarize the evolution of the stoichiometry for LSMO thin films. Figure 6 shows a general depth profile with typical values for metal occupations across the interface, in the film, and at the surface for an idealized 9-ML-thick LSMO film, based on the average experimental values in Fig. 5.

As we recently showed in the case of $LaAlO_3$ on TiO_2-terminated STO,[34] the valence state of the metal ion in

gesting that the gradual change of the stoichiometry across the interface is real.

We first focus on the Sr/La evolution. We use x' to denote the Sr content of a specific ML (and La$=1-x'$). For the three thicker films, we see x' changing from 0.85 to 0.4 as z increases from −0.5 to 1.5.[33] Above $z=1.5$, $x' \approx 0.3$, up to the topmost layer, where x' suddenly increases again and reaches a typical value of $x'=0.72$. This can be explained by preferential Sr segregation and will be discussed below. For the three thinnest films, the crossover to more La than Sr does not take place. A possible explanation could be that Sr segregation competes with the establishment of the "nominal" stoichiometry ($x'=0.35$). The disagreement for LSMO 3 between the stoichiometry determined by structural refinement and by RBS (Table II) is probably because with only 3 MLs, we are close to the limit of depth resolution of RBS.

We turn now to the transition of Ti to Mn. The two elements are very hard to distinguish with nonresonant x-ray techniques, as their atomic numbers only differ by 3. The fact that their occupation parameters have distinct values and show consistent trends can be explained with the accurate approximation of the atomic form factors in FIT and the reliability of our SXRD data. There are, however, limitations,

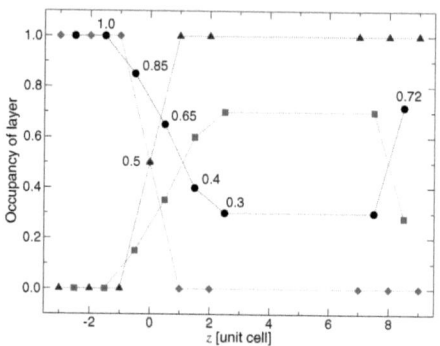

FIG. 6. (Color online) Evolution of the stoichiometries for an idealized film of 9-ML thickness. Sr=black •, La=red ■, Ti=green ♦, Mn=blue ▲. The dotted lines are guides to the eye.

the center of the oxygen octahedron is crucial to explain the dilation across the film-STO interface. Above 2.5 MLs (where we also observe the dilation maximum), we note that according to the general picture in Fig. 6, the La^{3+} occupation is established. We thus also have the highest concentration of (high-spin) Mn^{3+} that has an ionic radius 0.115 Å larger than Mn^{4+}.[35] Moreover, the films seem to be Mn rich at the interface, which would explain the increase of c at the interface and the subsequent decrease after the interfacial region. But the argument would still hold if instead of Mn a higher concentration of Ti was assumed, due to the additional Ti from the double layer, as the ionic radius of Ti^{3+} is 0.065 Å larger than that of Ti^{4+}. Either way, an enrichment of trivalent metal (Me^{3+}) species at the interface would help to explain the elongation of the out-of-plane lattice constant in the interface region.

Electrostatic considerations help to further corroborate any enrichment of the trivalent sites across the interface. The exchange of Sr and La cations lead to the formation of a dipole moment with the electric field and electric potential both being nonzero in the film. The trivalent species can minimize the electrostatic energy, if we require the electric field to be minimized and the electric potential to be zero at the film boundaries. We carried out such a minimization for the four interfacial MLs of LSMO 9 ($0 \leq z \leq 3$) and found that the enrichment of the trivalent species above the nominal interface is highest around $z=1$, leading to full occupation with Me^{3+}, with decreasing occupancy above and below. Note that the position of this maximum coincides with the highest gradient of the metal sites of LSMO 9 in Fig. 4. The minimization of the electrostatic energy thus supports the picture of a Me^{3+}-rich interface. However, a quantitative discussion of the evolution of the Ti^{3+}/Mn^{4+} ratio across the interface is beyond the accuracy of the used ionic radii,[35] in order to explain the experimentally observed lattice constants.

E. Strontium segregation

The film surface consistently shows a higher Sr content in the topmost ML than the film average. We attribute this to Sr segregation. The effect of segregation of the divalent site has been discussed before and was suggested to either behave exponentially[36] or occur exclusively in the outermost layer.[37,38] We cannot infer an exponential behavior from the data shown in Fig. 5, but do support a Sr-rich topmost unit cell. Moreover, the nature of the terminating layer at the film surface, either SrO (Ref. 37) or MnO_2 (Ref. 39), has been discussed in the literature. The COBRA results, on which the fits are based, demonstrate that the surface is terminated with a MnO_2 atomic layer. There is only a negligible amount of extra material above the nominal surface ML, in agreement with what one would expect for PLD growth on TiO_2-terminated STO substrates. Our FIT results support this: the addition of a partially occupied extra Sr/La-O atomic layer on top of the surface led to coverages at most of the order of the uncertainty of a laser burst, i.e., 7 to 8 % of a ML. Note also that the addition of such a sparsely occupied overlayer in the fit had no effect on the Sr enrichment of the complete (real) surface beneath it. In other words, the Sr enrichment is real and cannot be an artefact caused by a partially occupied overlayer.

Indeed, we can explain the observed Sr segregation by an intrinsic growth phenomenon, since LSMO 9 is a continuation from LSMO 6, i.e., both are grown on the same STO sample, and both films show a Sr enrichment only at the surface layer of the film. A surface segregation phenomenon suggests different enthalpies of the surface and film. We can write the total free energy F for a multicomponent system using a simple statistical model of segregation:[40]

$$F = \sum_i n_i^s g_i^s + n_i^f g_i^f - k_B T \ln \Omega, \quad (1)$$

where n_i^s and n_i^f are the number of surface and film atoms of type i, and g_i^s and g_i^f are the individual Gibbs free energies, respectively. k_B is the Boltzmann constant, T is the temperature, and Ω is the entropy due to the mixing of the compounds. The competition of minimizing the free energy terms of the individual components and maximizing the entropy causes the segregation.

We can derive an Arrhenius expression for the two component system, depending on the occupation x:

$$x_s = x_f \exp\left(\frac{-H}{k_B T}\right), \quad (2)$$

where $x_s = n_{Sr}^s / n_{La}^s$ and $x_f = n_{Sr}^f / n_{La}^f$ depend on the surface and film stoichiometry, respectively, and H is the segregation enthalpy. Using the values we summarized for the stoichiometry evolution in Fig. 6 ($x_s=2.57$, $x_f=0.429$), we obtain $H = -15$ kJ/mol (or -0.16 eV for a Sr site) for the segregation of Sr towards the film surface. Moreover, any kinetic barrier can be easily overcome by the energy transfer from the impinging particles of typically 5–25 eV kinetic energy during PLD growth.[41]

F. Bandwidth

At a fixed hole density x, the properties of the manganites are affected by distortions of the ideal cubic geometry, qualitatively expressed by the tolerance factor $\Gamma = (r_A + r_O)/[\sqrt{2}(r_B + r_O)]$. The transport properties depend on the overlap of the Mn sites with the O $2p$ orbitals, which in turn is determined by the Mn-O-Mn angle ϕ. $\phi < 180°$ result in a reduced electron hopping amplitude, proportional to $\cos \phi$.[42] In the case of LSMO, the tolerance factor $\Gamma = 0.98$ (for $x=0.35$), and $\phi < 180°$, and the tendency towards charge localization increases. Furthermore, in the Hubbard picture, electron hopping term t is not only dependent on the bond angle ϕ, but also on the length of the Mn-O bond as $1/(d_{Mn-O})^\alpha$, where $\alpha > 1$.[43] We therefore expect changes in T_C when ϕ and/or d_{Mn-O} change. This simple picture qualitatively explains the different Curie temperatures of $T_C = 370$ K, $T_C = 250$ K, and $T_C = 100$ K observed at a doping level of $x=0.3$ in $La_{1-x}Sr_xMnO_3$ ($\Gamma=0.98$), $La_{1-x}Ca_xMnO_3$ ($\Gamma=0.97$), and $Pr_{1-x}Ca_xMnO_3$ ($\Gamma=0.92$), respectively.[35,44]

The electronic bandwidth W is directly proportional to the electronic hopping term t in the Hubbard model, i.e., the ability of the Mn^{3+} e_g electrons to interact with the neighboring Mn^{4+} site via the O $2p$ orbital in order to develop the long-range ferromagnetic ordering in these materials. As a rule of thumb, the transition temperature T_C from a paramagnetic-insulating to a ferromagnetic conducting state increases with larger W. Note that this is usually accompanied by a reduction of the magnetoresistive effect.

The bandwidth W is influenced most by two relatively easily accessible structural quantities: the angle ϕ of a Mn-O-Mn bond and the Mn-O bond length d_{Mn-O}. For perovskite structures, W is a straightforward result of the tight-binding approximation. Empirically, the dependence of W on the bond angle and bond distance is given by[45]

$$W \propto \frac{\cos\omega}{(d_{Mn-O})^{3.5}}, \quad (3)$$

where $\omega=(\pi-\phi)/2$. Note that for calculations, average values for ϕ and d_{Mn-O} are usually taken, as one is interested in W for the complete layer of material. In the bulk, $\phi=166°$,[44] whereas the films of this work had bond angles $162<\phi<176°$. Moreover, as we pointed out in Sec. III C, the oxygen positions are less accurately determined compared to the metallic sites. Thus a modification of Eq. (3) probably leads to more reliable results under the assumption that $\sin\phi/2 \approx 1$:

$$W \propto \frac{1}{(d_{Mn-Mn}/2)^{3.5}}, \quad (4)$$

where $d_{Mn-Mn}/2$ is taken as the estimate for d_{Mn-O}.

Using Eq. (4), the refined atomic positions were taken to calculate the electronic bandwidth W for different film thicknesses, both ML for ML [Fig. 7(a)], and $\bar{W}$, averaged and weighted for different Mn occupations [Fig. 7(b)]. For comparison, W was calculated using the bulk LSMO lattice constant.

For the three thinnest films in Fig. 7(a), W is either significantly below the bulk value (LSMO 1) or rapidly decreasing (LSMO 2 and 3). LSMO 3 indeed had a resistance too high to be measured using our four-point setup. LSMO 4 had bandwidths comparable to that of the bulk. However, we could not measure any electrical conductance across LSMO 4. Whether this is due to insufficient experimental sensitivity or because this film is in fact insulating cannot be assessed. For the two thickest films, W is very close to the bulk value. Additionally, the averaged bandwidth $\bar{W}$ in Fig. 7(b) shows that for the thinner films (LSMO 1–4), the bandwidth is quite different from the bulk value. On the other hand, $\bar{W}$ for the two thickest films is very close to the bulk value. This suggests that magnetoresistive behavior becomes established between 6 and 9 MLs. The onset of magnetoresistive behavior of films of a thickness of 9 or more MLs can therefore be expected, and was indeed experimentally found in this work.

It is important to note that the mostly insulating properties of the surface when measuring the electrical resistance do not exclude the existence of magnetoresistance, as can be seen in Fig. 2 and in other experimental observations.[7] We

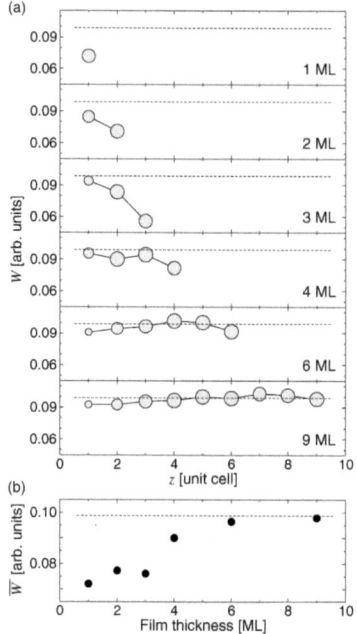

FIG. 7. Electronic bandwidth according to Eq. (4) for different film thicknesses based on the refined structure data: (a) W as a function of z calculated ML for ML with the size of the circles representing the weight of this particular data point; or (b) averaged and weighted for each film thickness. The dashed lines indicate bulk LSMO.

thus can confirm that the presence of an electrical dead layer does not imply an absence of magnetoresistance.

Finally, we note that the last data point of W in Fig. 7(a) is always somewhat lower compared to the next to last point. A possible explanation could involve the tendency of Sr to segregate towards the surface. This leads to a lower concentration of e_g electrons (and accordingly a higher hole concentration), which is reflected in the lower W value at the film surface. This might have important consequences for other experimental techniques probing mainly the surface such as x-ray photoelectron spectroscopy, and helps to explain angle-resolved photoemission data on this system.[46]

IV. CONCLUSIONS

We grew $La_{1-x}Sr_xMnO_3$ ($x=0.35$) thin films monolayer by monolayer on $SrTiO_3(001)$ with thicknesses of 1, 2, 3, 4, 6, and 9 MLs using pulsed laser deposition. We recorded large sets of structure factors for each film *in situ* by surface x-ray diffraction. The structures were analyzed using the direct method technique COBRA and subsequent refinement of the atomic positions.

We observe an inherent dilation of the substrate-film interface perpendicular to the surface. The atomic positions of the substrate-film interface are well defined, indicative of low roughness. However, the transition from substrate to film is not abrupt: The stoichiometry changes over approximately three unit cells. The MnO_2-terminated film surfaces are Sr enriched in the topmost monolayer, due to a segregation process. This enthalpy is estimated to be of the order of −15 kJ/mol (or −0.16 eV/Sr). Using the refined atomic positions, we calculated the electronic bandwidths for comparison with bulk LSMO. This led to the suggestion of an onset of magnetoresistance of nine or more MLs. The resistivity measurements for the 9-ML-thick LSMO film indeed confirm magnetoresistance, but they also indicate the presence of an electrically insulating dead layer, in agreement with other experimental observations for such very thin films.

ACKNOWLEDGMENTS

We are indebted to E. Dagotto for his invaluable comments on the manuscript. Fruitful discussions with O. Bunk and J. Krempasky are gratefully acknowledged. We thank S. Weyeneth from the University of Zürich, Switzerland, for his assistance in the resistivity measurements and his help in interpretation of the results. Support of this work by the Schweizerischer Nationalfonds zur Förderung der wissenschaftlichen Forschung and the staff of the Swiss Light Source is gratefully acknowledged. This work was partly performed at the Swiss Light Source, Paul Scherrer Institut. Work at the University of Michigan was supported in part by U.S. Department of Energy Grant No. DE-FG02-06ER46273 and by U.S. National Science Foundation Physics Frontier Center Grant No. PHY-0114336.

---

*philip.willmott@psi.ch

[1] G. H. Jonker and J. H. van Santen, Physica (Amsterdam) **16**, 337 (1950).

[2] S. Jin, T. H. Tiefel, M. McCormack, R. A. Fastnacht, R. Ramesh, and L. H. Chen, Science **264**, 413 (1994).

[3] P. Schiffer, A. P. Ramirez, W. Bao, and S.-W. Cheong, Phys. Rev. Lett. **75**, 3336 (1995).

[4] A. Urushibara, Y. Moritomo, T. Arima, A. Asamitsu, G. Kido, and Y. Tokura, Phys. Rev. B **51**, 14103 (1995).

[5] C. Zener, Phys. Rev. **82**, 403 (1951).

[6] A. J. Millis, P. B. Littlewood, and B. I. Shraiman, Phys. Rev. Lett. **74**, 5144 (1995).

[7] J.-H. Liao, T.-B. Wu, S.-T. Ho, Y.-T. Chen, S.-U. Jen, and Y.-D. Yao, J. Phys. D **40**, 4586 (2007).

[8] J. Z. Sun, D. W. Abraham, R. A. Rao, and C. B. Eom, Appl. Phys. Lett. **74**, 3017 (1999).

[9] R. P. Borges, W. Guichard, J. G. Lunney, J. M. D. Coey, and F. Ott, J. Appl. Phys. **89**, 3868 (2001).

[10] P. R. Willmott et al., Appl. Surf. Sci. **247**, 188 (2005).

[11] G. Koster, B. L. Kropman, G. J. H. M. Rijnders, D. H. A. Blank, and H. Rogalla, Appl. Phys. Lett. **73**, 2920 (1998).

[12] R. Herger, P. R. Willmott, O. Bunk, C. M. Schlepütz, B. D. Patterson, B. Delley, Phys. Rev. Lett. **98**, 076102 (2007).

[13] P. R. Willmott, R. Herger, M. C. Falub, L. Patthey, M. Döbeli, C. V. Falub, M. Shi, and M. Schneider, Appl. Phys. A: Mater. Sci. Process. **79**, 1199 (2004).

[14] P. R. Willmott, R. Herger, C. M. Schlepütz, D. Martoccia, and B. D. Patterson, Phys. Rev. Lett. **96**, 176102 (2006).

[15] B. D. Patterson et al., Nucl. Instrum. Methods Phys. Res. A **540**, 42 (2005).

[16] C. M. Schlepütz, R. Herger, P. R. Willmott, B. D. Patterson, O. Bunk, C. Brönnimann, B. Henrich, G. Hülsen, and E. F. Eikenberry, Acta Crystallogr., Sect. A: Found. Crystallogr. **61**, 418 (2005).

[17] R. Herger, P. R. Willmott, O. Bunk, C. M. Schlepütz, B. D. Patterson, B. Delley, V. L. Shneerson, P. F. Lyman, and D. K. Saldin, Phys. Rev. B **76**, 195435 (2007).

[18] M. Döbeli, R. M. Ender, V. Liechtenstein, and D. Vetterli, Nucl. Instrum. Methods Phys. Res. B **142**, 417 (1998).

[19] M. Döbeli, Nucl. Instrum. Methods Phys. Res. B **249**, 800 (2006).

[20] L. R. Doolittle, Nucl. Instrum. Methods Phys. Res. B **15**, 227 (1986).

[21] M. Björck and G. Andersson, J. Appl. Crystallogr. **40**, 1174 (2007).

[22] Y. Yacoby, M. Sowwan, E. Stern, J. Cross, D. Brewe, R. Pindak, J. Pitney, E. B. Dufresne, and R. Clarke, Physica B **336**, 39 (2003).

[23] D. D. Fong et al., Phys. Rev. B **71**, 144112 (2005).

[24] O. Bunk, Ph.D. thesis, University of Hamburg, Department of Physics, 1999, http://www.sub.uni-hamburg.de/opus/volltexte/1999/99/

[25] The R factor used here is defined as $R(|F|) = \dfrac{\sum_{i=1}^{N} ||F|_{i,exp} - |F|_{i,theor}|}{\sum_{i=1}^{N} |F|_{i,exp}}$, where $|F|$ is the amplitude of the structure factor.

[26] Note that PLD grows blocks of complete unit cells (i.e., ML by ML), and not layer by layer as, e.g., molecular beam epitaxy does.

[27] A. M. Haghiri-Gosnet, J. Wolfman, B. Mercey, C. Simon, P. Lecoeur, M. Korzenski, M. Hervieu, R. Desfeux, and G. Baldinozzi, J. Appl. Phys. **88**, 4257 (2000).

[28] We analyzed in addition the XRR curve of a 17-ML-thick film that confirmed the presence of an ≈20-Å-thick interface layer (although less dense). We thus can exclude that the two different optical densities shown in LSMO 6 and 9 are a result from the continuation of film growth on the STO sample 3.

[29] Note that in Fig. 2, we use the relationship $\rho = RGA/l$, where A is the film cross section, l is the distance between the electrodes, and $G = \pi/\ln(\sqrt{2})$ is a geometrical factor which accounts for the fact that the electric field lines do not run parallel through the film, but have a radial component. This factor of approximately 10 accounts for the apparent anomalously high resistivity compared to other literature values quoting the normal sheet resistivity.

[30] M. Stengel and N. A. Spaldin, Nature (London) **443**, 679 (2006).

[31] See EPAPS Document No. E-PRBMDO-77-063804 for tabulated atomic coordinates and occupation parameters in ASCII format. For more information on EPAPS, see http://www.aip.org/pubservs/epaps.html.

[32] K. Lonsdale, in *International Tables for X-Ray Crystallography*, edited by C. H. MacGillavry, G. D. Rieck, and K. Lonsdale (D. Reidel Publishing Company, Kluwer Academic Publishing, Dordrecht, 1985), Vol. III, Chap. 3.3.5, p. 232.

[33] Note that we also tried to fit the occupation parameters for the metallic sites for $z \leq -1$, but did not notice significant changes from nominal values of 0 and 1. In order to reduce the number of free parameters, we did not fit occupation parameters with $z \leq -1$.

[34] P. R. Willmott *et al.*, Phys. Rev. Lett. **99**, 155502 (2007).

[35] R. D. Shannon, Acta Crystallogr., Sect. A: Found. Crystallogr. **32**, 751 (1976).

[36] H. Dulli, P. A. Dowben, S. H. Liou, and E. W. Plummer, Phys. Rev. B **62**, R14629 (2000).

[37] R. Bertacco, J. P. Contour, A. Barthelemy, and J. Olivier, Surf. Sci. **511**, 366 (2002).

[38] J. W. Choi, J. D. Zhang, S. H. Liou, P. A. Dowben, and E. W. Plummer, Phys. Rev. B **59**, 13453 (1999).

[39] M. Izumi, Y. Konishi, T. Nishihara, S. Hayashi, M. Shinohara, M. Kawasaki, and Y. Tokura, Appl. Phys. Lett. **73**, 2497 (1998).

[40] S. Y. Liu and H. H. Kung, Surf. Sci. **110**, 504 (1981).

[41] P. R. Willmott and J. R. Huber, Rev. Mod. Phys. **72**, 315 (2000).

[42] E. Dagotto, T. Hotta, and A. Moreo, Phys. Rep. **344**, 1 (2001).

[43] W. A. Harrison, *Electronic Structure and the Properties of Solids: The Physics of the Chemical Bond* (Dover Publications, Inc., New York, 1989).

[44] H. Y. Hwang, S.-W. Cheong, P. G. Radaelli, M. Marezio, and B. Batlogg, Phys. Rev. Lett. **75**, 914 (1995).

[45] M. Medarde, J. Mesot, P. Lacorre, S. Rosenkranz, P. Fischer, and K. Gobrecht, Phys. Rev. B **52**, 9248 (1995).

[46] J. Krempaský *et al.* (unpublished).

Publications by R. Herger

[1] P. R. Willmott, R. Herger, and C. M. Schlepütz, *Multilayers, alloys, and complex profiles by pulsed laser deposition using a novel target geometry*, Thin Solid Films **453-54**, 436 (2004).

[2] P. R. Willmott, R. Herger, M. C. Falub, L. Patthey, M. Döbeli, C. V. Falub, M. Shi, and M. Schneider, *Pulsed laser deposition of atomically flat $La_{1-x}Sr_xMnO_3$ thin films using a novel target geometry*, Appl. Phys. A **79**, 1199 (2004).

[3] M. Shi, M. C. Falub, P. R. Willmott, J. Krempaský, R. Herger, K. Hricovini, and L. Patthey, *k-dependent electronic structure of the colossal magnetoresistive perovskite $La_{0.66}Sr_{0.34}MnO_3$*, Phys. Rev. B **70**, 140407(R) (2004).

[4] P. R. Willmott, C. M. Schlepütz, R. Herger, B. D. Patterson, K. Hassdenteufel, and W. Steurer, *In situ diffraction studies of the initial growth processes of textured icosahedral quasicrystalline thin films*, Phys. Rev. B **71**, 094203 (2005).

[5] P. R. Willmott, R. Herger, B. D. Patterson, and R. Windiks, *Experimental and theoretical study of the strong dependence of the microstructural properties of $Sr_xBa_{1-x}Nb_2O_6$ thin films as a function of their composition*, Phys. Rev. B **71**, 144114 (2005).

[6] C. M. Schlepütz, R. Herger, P. R. Willmott, B. D. Patterson, O. Bunk, C. Brönnimann, B. Henrich, G. Hülsen, and E. F. Eikenberry, *Improved data acquisition in grazing-incidence x-ray scattering experiments using a pixel detector*, Acta Crystallogr. Sect. A **61**, 418 (2005).

[7] P. R. Willmott, C. M. Schlepütz, B. D. Patterson, R. Herger, M. Lange, D. Meister, D. Maden, C. Brönnimann, E. F. Eikenberry, G. Hülsen, and A. Al-Adwan, *In situ studies of complex PLD-grown films using hard x-ray surface diffraction*, Appl. Surf. Sci. **247**, 188 (2005).

[8] P. R. Willmott, R. Herger, C. M. Schlepütz, D. Martoccia, and B. D. Patterson, *Energetic surface smoothing of complex metal-oxide thin films*, Phys. Rev. Lett. **96**, 176102 (2006).

[9] O. Bunk, M. Corso, D. Martoccia, R. Herger, P. R. Willmott, B. D. Patterson, J. Osterwalder, J. F. van der Veen, and T. Greber, *Surface x-ray diffraction study of boron-nitride nanomesh in air*, Surf. Sci. **601**, L7 (2007).

[10] R. Herger, P. R. Willmott, O. Bunk, C. M. Schlepütz, B. D. Patterson, and B. Delley, *Surface of strontium titanate*, Phys. Rev. Lett. **98**, 076102 (2007).

[11] P. R. Willmott, S. A. Pauli, R. Herger, C. M. Schlepütz, D. Martoccia, B. D. Patterson, B. Delley, R. Clarke, D. Kumah, C. Cionca, and Y. Yacoby, *Structural basis for the conducting interface between $LaAlO_3$ and $SrTiO_3$*, Phys. Rev. Lett. **99**, 155502 (2007).

[12] R. Herger, P. R. Willmott, O. Bunk, C. M. Schlepütz, B. D. Patterson, B. Delley, V. L. Shneerson, P. F. Lyman, and D. K. Saldin, *Surface structure of $SrTiO_3(001)$*, Phys. Rev. B **76**, 195435 (2007).

[13] S. A. Pauli, R. Herger, P. R. Willmott, E. V. Donev, J. Y. Suh, and R. F. J. Haglund, *X-ray diffraction studies of the growth of vanadium dioxide nanoparticles*, J. Appl. Phys. **102**, 073527 (2007).

[14] R. Herger, P. R. Willmott, C. M. Schlepütz, M. Björck, S. A. Pauli, D. Martoccia, B. D. Patterson, D. Kumah, R. Clarke, Y. Yacoby, and M. Döbeli, *Structure determination of monolayer-by-monolayer grown $La_{1-x}Sr_xMnO_3$ thin films and the onset of magnetoresistance*, Phys. Rev. B **77**, 085401 (2008).

[15] J. Krempaský, V. N. Strocov, L. Patthey, P. R. Willmott, R. Herger, M. Falub, P. Blaha, M. Hoesch, V. Petrov, M. C. Richter, O. Heckmann, and K. Hricovini, *Effects of three-dimensional band structure in angle- and spin-resolved photoemission from half-metallic $La_{2/3}Sr_{1/3}MnO_3$*, Phys. Rev. B **77**, 165120 (2008).

[16] M. Björck, C. M. Schlepütz, S. A. Pauli, D. Martoccia, R. Herger, and P. R. Willmott, *Atomic imaging in thin films with x-ray surface diffraction: introducing DCAF*, J. Phys. Condens. Matter **20**, 445006 (2008).

Die VDM Verlagsservicegesellschaft sucht für wissenschaftliche Verlage abgeschlossene und herausragende

Dissertationen, Habilitationen, Diplomarbeiten, Master Theses, Magisterarbeiten usw.

für die kostenlose Publikation als Fachbuch.

Sie verfügen über eine Arbeit, die hohen inhaltlichen und formalen Ansprüchen genügt, und haben Interesse an einer honorarvergüteten Publikation?

Dann senden Sie bitte erste Informationen über sich und Ihre Arbeit per Email an *info@vdm-vsg.de*.

Sie erhalten kurzfristig unser Feedback!

VDM Verlagsservicegesellschaft mbH
Dudweiler Landstr. 99   Telefon +49 681 3720 174
D - 66123 Saarbrücken   Fax   +49 681 3720 1749
www.vdm-vsg.de

Die VDM Verlagsservicegesellschaft mbH vertritt

Printed by Books on Demand GmbH, Norderstedt / Germany